# AQA
## GCSE chemistry

For AQA

Author

Philippa Gardom Hulme

# Contents

# How to use this book

**Welcome to your AQA GCSE Chemistry course.** This book has been specially written by experienced teachers and examiners to match the 2011 specification.

On these two pages you can see the types of pages you will find in this book, and the features on them. Everything in the book is designed to provide you with the support you need to help you prepare for your examinations and achieve your best.

## Unit openers

**Specification matching grid:** This shows you how the pages in the unit match to the exam specification for GCSE Chemistry, so you can track your progress through the unit as you learn.

**Why study this unit:** Here you can read about the reasons why the science you're about to learn is relevant to your everyday life.

**You should remember:** This list is a summary of the things you've already learnt that will come up again in this unit. Check through them in advance and see if there is anything that you need to recap on before you get started.

**Opener image:** Every unit starts with a picture and information on a new or interesting piece of science that relates to what you're about to learn.

## Main pages

**Learning objectives:** You can use these objectives to understand what you need to learn to prepare for your exams. Higher Tier only objectives appear in pink text.

**Key words:** These are the terms you need to understand for your exams. You can look for these words in the text in bold or check the glossary to see what they mean.

**Questions:** Use the questions on each spread to test yourself on what you've just read.

**Higher Tier content:** Anything marked in pink is for students taking the Higher Tier paper only. As you go through you can look at this material and attempt it to help you understand what is expected for the Higher Tier.

**Worked examples:** These help you understand how to use an equation or to work through a calculation. You can check back whenever you use the calculation in your work.

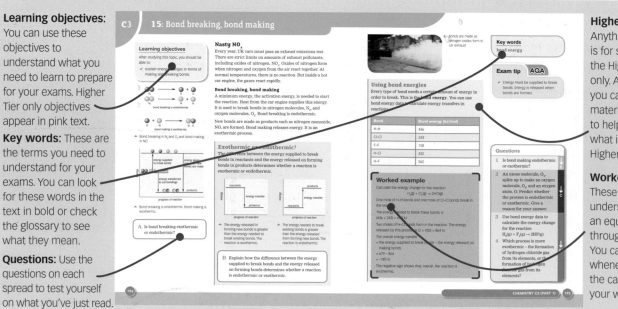

4

# Summary and exam-style questions

Every summary question at the end of a spread includes an indication of how hard it is. These indicators show which grade you are working towards. You can track your own progress by seeing which of the questions you can answer easily, and which you have difficulty with.

When you reach the end of a unit you can use the exam-style questions to test how well you know what you've just learnt. Each question has a grade band next to it.

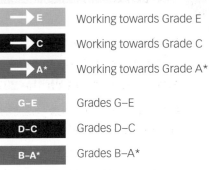

| | |
|---|---|
| → E | Working towards Grade E |
| → C | Working towards Grade C |
| → A* | Working towards Grade A* |
| G–E | Grades G–E |
| D–C | Grades D–C |
| B–A* | Grades B–A* |

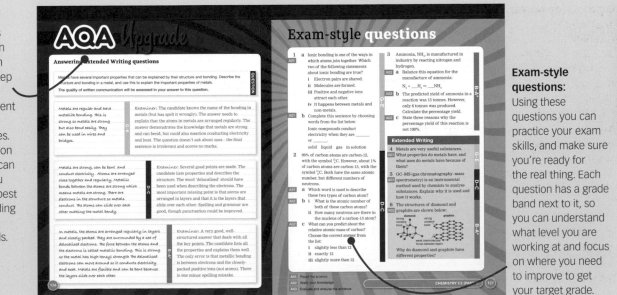

**Revision checklist:** This is a summary of the main ideas in the unit. You can use it as a starting point for revision, to check that you know about the big ideas covered.

**Visual summary:** Another way to start revision is to use a visual summary, linking ideas together in groups so you can see how one topic relates to another. You can use this page as a start for your own summary.

**Upgrade:** Upgrade takes you through an exam question in a step-by-step way, showing you why different answers get different grades. Using the tips on the page you can make sure you achieve your best by understanding what each question needs.

**Exam-style questions:** Using these questions you can practice your exam skills, and make sure you're ready for the real thing. Each question has a grade band next to it, so you can understand what level you are working at and focus on where you need to improve to get your target grade.

# Routes and assessment

## Matching your course

The units in this book have been written to match the specification for **AQA GCSE Chemistry**.

In the diagram below you can see that the units and part units can be used to study either for **GCSE Chemistry** or as part of **GCSE Science** and **GCSE Additional Science** courses.

|  | GCSE Biology | GCSE Chemistry | GCSE Physics |
|---|---|---|---|
| **GCSE Science** | B1 (Part 1) | C1 (Part 1) | P1 (Part 1) |
|  | B1 (Part 2) | C1 (Part 2) | P1 (Part 2) |
| **GCSE Additional Science** | B2 (Part 1) | C2 (Part 1) | P2 (Part 1) |
|  | B2 (Part 2) | C2 (Part 2) | P2 (Part 2) |
|  | B3 (Part 1) | C3 (Part 1) | P3 (Part 1) |
|  | B3 (Part 2) | C3 (Part 2) | P3 (Part 2) |

## GCSE Chemistry assessment

The units in this book are broken into two parts to match the different types of exam paper on offer. The diagram below shows you what is included in each exam paper. It also shows you how much of your final mark you will be working towards in each paper.

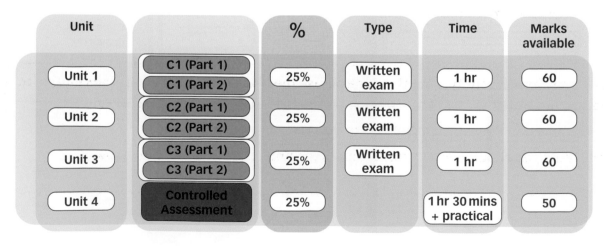

| Unit | | % | Type | Time | Marks available |
|---|---|---|---|---|---|
| Unit 1 | C1 (Part 1) C1 (Part 2) | 25% | Written exam | 1 hr | 60 |
| Unit 2 | C2 (Part 1) C2 (Part 2) | 25% | Written exam | 1 hr | 60 |
| Unit 3 | C3 (Part 1) C3 (Part 2) | 25% | Written exam | 1 hr | 60 |
| Unit 4 | Controlled Assessment | 25% | | 1 hr 30 mins + practical | 50 |

# Understanding exam questions

When you read the questions in your exam papers you should make sure you know what kind of answer you are being asked for. The list below explains some of the common words you will see used in exam questions. Make sure you know what each word means. Always read the question thoroughly, even if you recognise the word used.

### Calculate
Work out your answer by using a calculation. You can use your calculator to help you. You may need to use an equation; check whether one has been provided for you in the paper. The question will say if your working must be shown.

### Describe
Write a detailed answer that covers what happens, when it happens, and where it happens. The question will let you know how much of the topic to cover. Talk about facts and characteristics. (Hint: don't confuse with 'Explain')

### Explain
You will be asked how or why something happens. Write a detailed answer that covers how and why a thing happens. Talk about mechanisms and reasons. (Hint: don't confuse with 'Describe')

### Evaluate
You will be given some facts, data or other information. Write about the data or facts and provide your own conclusion or opinion on them.

### Outline
Give only the key facts of the topic. You may need to set out the steps of a procedure or process – make sure you write down the steps in the correct order.

### Show
Write down the details, steps or calculations needed to prove an answer that you have been given.

### Suggest
Think about what you've learnt in your science lessons and apply it to a new situation or a context. You may not know the answer. Use what you have learnt to suggest sensible answers to the question.

### Write down
Give a short answer, without a supporting argument.

### Top tips
Always read exam questions carefully, even if you recognise the word used. Look at the information in the question and the number of answer lines to see how much detail the examiner is looking for.

You can use bullet points or a diagram if it helps your answer.

If a number needs units you should include them, unless the units are already given on the answer line.

# Controlled Assessment in GCSE Chemistry

As part of the assessment for your GCSE Chemistry course, you will undertake a Controlled Assessment task.

## What is Controlled Assessment?

Controlled Assessment has taken the place of coursework for the new 2011 GCSE Science specifications. The main difference between coursework and Controlled Assessment is that you will be supervised by your teacher when you carry out your Controlled Assessment task.

## What will my Controlled Assessment task look like?

Your Controlled Assessment task will be made up of four sections. These four sections make up an investigation, with each section looking at a different part of the scientific process.

|  | What will I need to do? | How many marks are available? |
|---|---|---|
| **Research** | • Independently develop your own hypothesis.<br>• Research two methods for carrying out an experiment to test your hypothesis.<br>• Prepare a table to record your results.<br>• Carry out a risk assessment. | |
| **Section 1** | • Answer questions relating to your own research. | 20 marks |
| **Practical investigation** | • Carry out your own experiment and record and analyse your results. | |
| **Section 2** | • Answer questions relating to the experiment you carried out.<br>• Select appropriate data from data supplied by AQA and use it to analyse and compare with your hypothesis.<br>• Suggest how ideas from your investigation could be used in a new context. | 30 marks |
| | | **Total  50 marks** |

## How do I prepare for my Controlled Assessment?

Throughout your course you will learn how to carry out investigations in a scientific way, and how to analyse and compare data properly.

On the next three pages there are Controlled Assessment-style questions matched to the content in C1, C2, and C3. You can use them to test yourself, and to find out which areas you want to practise more before you take the Controlled Assessment task itself.

# C1 Controlled Assessment-style questions

**Hypothesis:** It is suggested that there is a link between the position of a metal in the reactivity series, and the ease of decomposition of its carbonate.

Download the Research Notes and Data Sheet for C1 from **www.oxfordsecondary.co.uk/aqacasestudies**.

## Research

*Record your findings in the **Research Notes table**.*

1. Research two different methods that could be used to test the hypothesis.
2. Find out how the results of the investigation might be useful in determining how to extract metals from ores that contain metal carbonates.

## Section 1      Total 20 marks

*Use your research findings to answer these questions.*

1. (a) Name two sources that you used for your research.
   (b) Which of these sources did you find more useful, and why? [3]
2. (a) Identify one control variable.
   (b) Briefly describe a preliminary investigation to find a suitable value for this variable, and explain how the results of this work will help you decide on the best value for it. [3]
3. *In this question you will be assessed on using good English, organising information clearly, and using specialist terms where appropriate.* Describe how to carry out an investigation to test the hypothesis. Include the equipment needed and how to use it, the measurements to make, how to make it a fair test, and a risk assessment. [9]
4. Use your research to outline another possible method, and explain why you did not choose it. [3]
5. Draw a table to record data from the investigation. You may use ICT if you wish. [2]

## Section 2      Total 30 marks

*Use the **Data Sheet** to answer these questions.*

1. (a) State the independent and dependent variables, and one control variable. [3]
   (b) The smallest scale division on a measuring instrument is called its resolution. What was the resolution of the instrument you used, and was this resolution suitable for your experiment? [3]
   (c) Display the **Group A data** on a bar chart or line graph. *This data has been provided for you to use instead of data that you would gather yourself.* [4]
   (d) Does the **Group A data** support the hypothesis? Explain how. [3]
   (e) Describe the similarities and differences between the **Group A data** and the **Group B data**. Suggest one reason why the results of the two groups may be different. *The **Group B data** has been provided for you to use instead of data that would be gathered by others in your class.* [3]
   (f) Suggest how your method might have helped Group A to achieve results that show a clear pattern. [3]
2. (a) Sketch a bar chart or line graph of the results in **Case study 1**. [2]
   (b) Explain to what extent the data from **Case studies 1–3** support the hypothesis. [3]
   (c) Use **Case study 4** to describe the relationship between the position of a Group 1 metal in the periodic table, and the ease of decomposition of its nitrate. Explain how well the data supports your answer. [3]
3. The context of this investigation (the topic it relates to) is determining how to extract metals from ores that contain metal carbonates. Describe how your results may be useful in this context. [3]

**Overview:** A change in temperature may cause a change in the rate of a reaction. **You must develop your own hypothesis to test.** You will be provided with magnesium ribbon, dilute hydrochloric acid, a thermometer, heating apparatus, and common laboratory glassware.

> Download the Research Notes and Data Sheet for C2 from **www.oxfordsecondary.co.uk/ aqacasestudies**.

## Research

*Record your findings in the **Research Notes table**.*

1. Research two methods to find out whether temperature affects the reaction rate of acids.
2. Find out how the investigation results might be useful in choosing the best temperature at which to use a new limescale remover.

## Section 1                                Total 20 marks

*Use your research findings to answer these questions.*

1. **(a)** Name the two most useful sources that you used for your research.
   **(b)** Explain why these sources were the most useful.                                         [3]
2. Write a hypothesis about how temperature might affect reaction rate. Use your research findings to explain why you made this hypothesis.                                         [3]
3. Describe how to carry out an investigation to test your hypothesis. Include the equipment needed and how to use it, the measurements to make, how to make it a fair test, and a risk assessment.                                         [9]
4. Use your research to outline another possible method, and explain why you did not choose it.                                         [3]
5. Draw a table to record data from the investigation. You may use ICT if you wish. [2]

## Section 2                                Total 30 marks

*Use the **Data Sheet** to answer these questions.*

1. Display the **Group A data** on a graph. *This data has been provided for you to use instead of data that you would gather yourself.*                                         [4]
2. **(a)** What conclusion can you draw from the **Group A data** about a link between temperature and reaction rate? Use any pattern you can see in the **Group A data** and quote figures from it to support your answer.                                         [3]
   **(b) (i)** Compare the **Group A** and **Group B data**. Do you think the **Group A data** is reproducible? Explain why. *The Group B data has been provided for you to use instead of data that would be gathered by others in your class.*   [3]
   **(ii)** Explain how you could use the repeated results from Group B to obtain a more accurate answer.                                         [3]
   **(c)** Look at the **Group A data**. Are there any anomalous results? Quote from the data to explain your answer.                                         [3]
3. **(a)** Sketch a graph of the results in **Case study 1**.                                         [2]
   **(b)** Explain to what extent the data from **Case studies 1–3** support or contradict your hypothesis.                                         [3]
   **(c)** Compare the **Group A data** to the data shown on the **Case study 4** graph. Explain how far the **Case study 4** data supports or contradicts your hypothesis.                                         [3]
4. A company is developing a new limescale remover. It needs to choose the best temperature for its use. A chemist develops a hypothesis that the higher the temperature, the faster the reaction, and the quicker limescale is removed.
   **(a)** Does the **Group A data** support or contradict this hypothesis? Quote figures from the data to explain your answer.                                         [3]
   **(b)** Show how ideas from the **Group A data** and the **Case studies** could be used by the company.                                         [3]

# C3 Controlled Assessment-style questions

**Overview:** The position of a metal in the reactivity series may be linked to the amount of heat energy released when it reacts with copper sulfate solution. **You must develop your own hypothesis to test.** You will be provided with zinc, iron, tin, and lead metals, copper sulfate solution, a thermometer, and common laboratory glassware.

Download the Research Notes and Data Sheet for C3 from **www.oxfordsecondary.co.uk/ aqacasestudies**.

## Research

*Record your findings in the **Research Notes table**.*

1. Research two methods to explore the link between the position of a metal in the reactivity series and the energy released when the metal reacts with copper sulfate solution.
2. Find out how the results of the investigation might help in choosing the best metal to add to copper sulfate solution to heat a ready meal.

## Section 1                Total 20 marks

*Use your research findings to answer these questions.*

1. (a) Name the two most useful sources that you used for your research.
   (b) Explain why these sources were the most useful.                [3]
2. Write a hypothesis about how the position of a metal in the reactivity series affects the energy released when it reacts with copper sulfate solution. Use your research findings to explain why you've made this hypothesis.    [3]
3. Describe how to carry out an investigation to test your hypothesis. Include the equipment needed and how to use it, the measurements to make, how to make it a fair test, and a risk assessment.    [9]
4. Use your research to outline another possible method, and explain why it was not chosen. [3]
5. Draw a table to record data from the investigation. You may use ICT if you wish. [2]

## Section 2                Total 30 marks

1. Display the **Group A data** on a graph or bar chart. *This data has been provided for you to use instead of data that you would gather yourself.*    [4]
2. (a) What conclusion can you draw from the **Group A data** about a link between position in reactivity series and energy released? Use any pattern you can see in the **Group A data** and quote figures.    [3]
   (b) (i) Compare the **Group A** and **Group B data**. Do you think the **Group A data** is reproducible? Explain why. *The **Group B data** has been provided for you to use instead of data that would be gathered by others in your class.*    [3]
       (ii) Explain how you could use the repeated results from Group B to obtain a more accurate answer.    [3]
   (c) Look at the **Group A data**. Are there any anomalous results? Quote from the data. [3]
3. (a) Sketch a bar chart of the results in **Case study 1**.    [2]
   (b) Explain to what extent the data from **Case studies 1–3** support or contradict your hypothesis.    [3]
   (c) Compare the **Group A data** to the **Case study 4 data**. Explain to what extent the **Case study 4** data supports or contradicts your hypothesis.    [3]
4. A company developing a new ready meal warmer wants to choose the most efficient pair of substances. A chemist develops a hypothesis that the higher a metal is in the reactivity series, the more energy is released when it reacts with copper sulfate solution.
   (a) Does the **Group A data** support or contradict this hypothesis?    [3]
   (b) Explain how the company could use ideas from the **Group A data** and the case studies, and suggest why it might decide against using copper sulfate solution to heat the ready meal.    [3]

# C1 Part 1

# Atoms, rocks, metals, and fuels

## Why study this unit?

The Earth's crust provides many resources for us. Extracting metals, rocks, and oil products can damage the environment and create dangerous waste.

In this unit you will discover how to extract metals from the Earth's crust, and how to make metal alloys with perfect properties for particular purposes. You will learn where limestone comes from, and about the chemical reactions that produce cement and concrete from this vital raw material. You will find out about fossil fuels too, and discover how the compounds of crude oil are separated to make petrol, diesel, and jet fuel.

### You should remember

1 Everything is made up of tiny particles called atoms.

2 In chemical reactions, atoms are rearranged. Energy is given out or taken in as new products are made.

3 There are patterns in the chemical reactions of substances such as metals.

4 The way we use materials depends on their properties.

5 There are patterns in the properties of materials such as metals.

6 Humans extract many useful materials, such as metals and oil, from the Earth's crust.

The 92-storey Trump International Hotel and Tower in Chicago is the tallest residential building in the United States. The tower was made from about 330 000 tonnes of concrete, 22 500 tonnes of steel, and many tonnes of glass. During construction, a huge pump forced liquid concrete to the upper layers of the tower at a rate of over 2700 kg per minute.

## Key words

metal, element, atom, symbol, periodic table, group

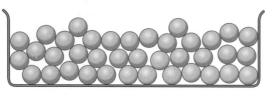

▲ A model of the atoms in liquid mercury. Each circle represents one mercury atom.

## Did you know...?

You are made up of atoms of elements, too. A 50 kg person contains 32.5 kg of oxygen atoms, 9 kg of carbon atoms, 5 kg of hydrogen atoms, and smaller amounts of many other elements.

A  What is an element?

B  What is an atom?

## Killer cargo

It's 1810. A ship takes on board a cargo of mercury. Within three weeks, many crew members are ill. Most are dribbling uncontrollably. Many have mouth ulcers and bowel complaints. Some are suffering more badly. Their faces are so swollen that their eyes will not open. Their tongues are so swollen that they can hardly breathe. Three sailors die. So do all the sheep, pigs, and cats on the boat.

These sailors and animals were suffering from mercury poisoning. The leather of the mercury containers had rotted. Liquid mercury flowed all over the ship. Sailors breathed in mercury vapour and absorbed the **metal** through their skin.

▲ Mercury is the only metal which is liquid at room temperature

## Using mercury

Mercury is not all bad. Its vapour is a vital part of low energy light bulbs – each bulb contains about five milligrams of the metal. Mercury saves lives in tip-over switches in electric heaters, too. When the heater is on, electricity flows through the mercury in the switch. If the heater falls over, the mercury moves. The circuit is broken, so the heater switches off and is unlikely to start a fire.

## Elements and atoms

Mercury is an **element**. It is made of just one sort of atom. An **atom** is the smallest part of an element that can exist. Atoms are tiny. The diameter of one atom is about 0.000 000 01 cm. If you could place one hundred million atoms side by side, they would stretch one centimetre.

Mercury is not the only element. In total, there are about 100 elements, each with its own type of atom. You cannot split elements into simpler substances. Everything in the world is made from the atoms of one or more of these 100 or so elements.

## Symbols for elements

Each element has its own **symbol**. The symbol for mercury is Hg. This comes from its Latin name, *hydrargyrum*. The symbol O represents one atom of oxygen. The list below shows the symbols of some other elements:

- nitrogen – N
- neon – Ne
- neptunium – Np
- sodium – Na.

**C** Name an element whose symbol is the first letter of its name.

**D** Suggest a reason for the symbols of neon and neptunium being two letters long.

## The periodic table

All the elements are listed in the **periodic table**. The vertical columns are called **groups**. The elements in a group have similar properties.

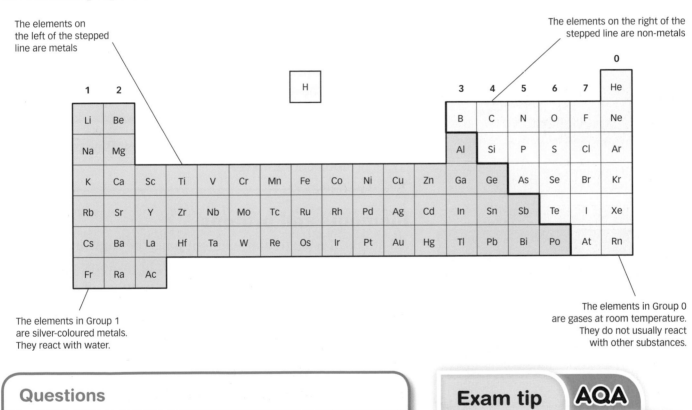

The elements on the left of the stepped line are metals

The elements on the right of the stepped line are non-metals

The elements in Group 1 are silver-coloured metals. They react with water.

The elements in Group 0 are gases at room temperature. They do not usually react with other substances.

## Questions

1 Give the name and symbol of one element in Group 1 of the periodic table.

2 State what all the elements in Group 0 have in common.

3 How many types of atom are there? Explain how you worked out your answer.

E

C

## Exam tip AQA

✓ Remember that everything is made up of atoms of about 100 elements.

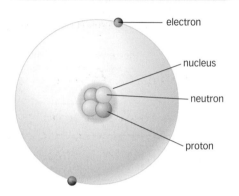

▲ Protons and electrons are electrically charged. Neutrons have no electrical charge. They are neutral.

| Name of particle | Charge |
|------------------|--------|
| proton | +1 |
| neutron | 0 |
| electron | −1 |

Electronic structure ▶ of hydrogen

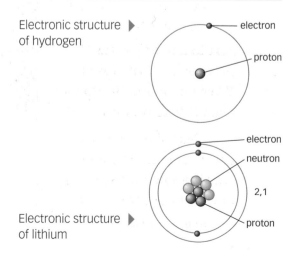

Electronic structure ▶ of lithium

## Exploring atoms

For many years, scientists believed that atoms were solid particles, like miniature snooker balls. But by the late 1800s, some scientists were beginning to doubt this idea. Scientists such as J. J. Thomson and Ernest Rutherford wanted to work out what atoms are really like. They carried out experiments to gather evidence.

## Inside atoms

We now know that an atom is mainly empty space. At its centre is a **nucleus**. The nucleus is made up of tiny particles called **protons** and **neutrons**. Outside the nucleus are even tinier particles, called **electrons**.

Overall, an atom has no electrical charge. This is because it has equal numbers of positive protons and negative electrons. An oxygen atom, for example, has eight protons and eight electrons. So the atom has no overall electrical charge.

> **A** A nitrogen atom has seven electrons. How many protons does it have?

## Elements and protons

All atoms of a particular element have the same number of protons. So every oxygen atom has eight protons, and every nitrogen atom has seven protons. The number of protons in an atom of an element is its **atomic number**. Atoms of different elements have different numbers of protons. The sum of the protons and neutrons in an atom is its **mass number**.

## Arranging electrons

Electrons are arranged in **energy levels**. Each electron in an atom is in a particular energy level. Electrons fill the lowest energy levels first.

Hydrogen has just one electron. It occupies the lowest available energy level.

Lithium has three electrons. Two electrons fill up the lowest energy level. The other electron goes in the next energy level.

The electrons in sodium and potassium are arranged like this.

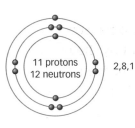

▲ Electronic structure of sodium

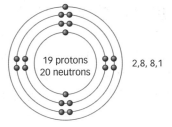

▲ Electronic structure of potassium

In all atoms, the lowest energy level can hold a maximum of two electrons. The next energy level holds up to eight electrons.

> **B** Neon has 10 electrons and argon has 18. Draw the electrons in an atom of neon and in an atom of argon.

## Electron arrangements and the periodic table

The elements lithium, sodium, and potassium are in Group 1 of the periodic table. The atoms of each of these elements have just one electron in their highest energy level.

Similar electron arrangements give Group 1 elements similar properties. Each reacts vigorously with water to make hydrogen gas and a metal hydroxide. For example

$$\text{potassium} + \text{water} \rightarrow \text{potassium hydroxide} + \text{hydrogen}$$

◀ Potassium reacts vigorously with water

Group 1 metals also have similar reactions when burned in air. They all react vigorously with oxygen to make a metal oxide.

$$\text{sodium} + \text{oxygen} \rightarrow \text{sodium oxide}$$

The elements neon, argon, and krypton are **noble gases**. They are in Group 0 of the periodic table. Their atoms have eight electrons in the highest energy level. Helium is in Group 0 too. Its atoms have two electrons in its highest energy level. These electron arrangements are very stable. So the noble gases are **unreactive**.

**Key words**

nucleus, proton, neutron, electron, atomic number, mass number, energy level, noble gas, unreactive

**Did you know...?**

If you could make an atom as big as a football stadium, its nucleus would be as small as a grain of sand.

**Exam tip**

✔ It might help to think of electron energy levels as shells. The lowest energy level is the innermost shell.

**Questions**

1 An atom of neon has ten protons. How many electrons does it have?

2 Draw diagrams to show the electron arrangements of carbon (six electrons), oxygen (eight electrons), and neon (ten electrons).

3 An atom of argon has 18 protons and 22 neutrons. Work out its atomic number and its mass number.

4 Use ideas about electron arrangement to explain why the elements in Group 0 have similar properties.

### Learning objectives

After studying this topic, you should be able to:

✔ explain the differences between a compound and an element

✔ explain ionic and covalent bonding

▲ If you eat too much salt you are more likely to get heart disease

▲ Sodium

▲ Chlorine

▲ Sodium chloride

## Vital element

Sodium is an element. It is vital to life. You need sodium to make your heart beat properly and to help your nerves transmit messages.

Of course, you can't eat pure sodium – the metal reacts violently with water. Instead you need to eat substances that contain sodium, such as sodium chloride, or salt.

## Compounds

Sodium chloride is a **compound.** Compounds are made up of two or more elements, strongly joined together.

The properties of compounds are different from the properties of the elements from which they are made. For example:

- Sodium is a solid silver-coloured metal at room temperature. It conducts electricity.
- Chlorine is a green gas. It is a non-metal. It does not conduct electricity.
- Sodium chloride is a white solid. It does not conduct electricity when solid.

> A  What is a compound?
>
> B  Describe two ways in which the compound sodium chloride is different from each of the elements sodium and chlorine.

## Holding compounds together
### Ionic bonds

Sodium chloride is made up of a metal joined to a non-metal. When sodium chloride forms from its elements, each sodium atom transfers one of its electrons to a chlorine atom.

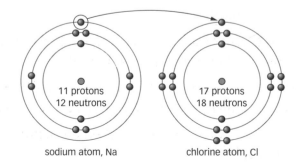

sodium atom, Na          chlorine atom, Cl

Each atom now has eight electrons in its highest energy level. These electron arrangements are very stable.

- The sodium atom now has 10 electrons for its 11 protons. Overall, it has a charge of +1.
- The chlorine atom now has 18 electrons and 17 protons. Overall, it has a charge of –1.

Charged atoms are called **ions**. Here, we have formed a positive sodium ion and a negative chloride ion.

Sodium chloride is an **ionic compound**. In ionic compounds, the positive and negative ions are strongly attracted to each other. The forces of attraction between positive and negative ions are called **ionic bonds**. A sodium chloride crystal consists of millions of sodium and chloride ions, all held together in a regular pattern by ionic bonds.

Ionic bonds only exist in compounds made up of metals and non-metals.

## Covalent bonds

The compound carbon dioxide is made up of two non-metals. So it cannot contain ionic bonds – there are no metal atoms to give away electrons.

Instead, its atoms are joined together in groups of three. Each group has one carbon atom and two oxygen atoms. The atoms share electrons between them to form **covalent bonds**.

Other compounds made up of non-metals only are held together by covalent bonds. The atoms in gases such as oxygen, $O_2$, and chlorine, $Cl_2$, also share electrons to form covalent bonds.

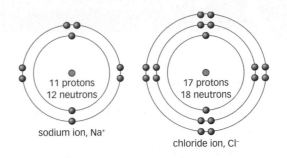

11 protons
12 neutrons

sodium ion, Na⁺

17 protons
18 neutrons

chloride ion, Cl⁻

---

**C** What is an ionic bond?

**D** Potassium chloride is held together by ionic bonds. Predict the charges on the potassium and chloride ions. Explain your predictions.

---

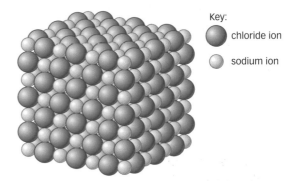

Key:
- chloride ion
- sodium ion

▲ Sodium ions and chloride ions are arranged like this

---

## Key words

compound, ion, ionic compound, ionic bond, covalent bond

---

## Questions

1 How many elements join together to make the compound sodium chloride?

2 Use ideas about atoms to explain the difference between an element and a compound.

3 Explain how sodium and chloride ions are formed when sodium chloride is made from its elements.

4 Explain how the atoms in a carbon monoxide molecule are held together.

↓ E

↓ C

↓ A*

---

**Exam tip**

✔ Remember – compounds made up of a metal and a non-metal are held together by ionic bonds. Compounds made up of non-metals only are held together by covalent bonds.

## Learning objectives

After studying this topic, you should be able to:

- ✔ understand what happens in chemical reactions
- ✔ write word equations to summarise reactions
- ✔ interpret and write chemical formulae

## Key words

chemical reaction, reactant, product, formula

**A** Describe two features of chemical reactions.

**B** Carbon burns in oxygen to make carbon dioxide. Write a word equation for the reaction.

**C** Methane burns in oxygen to make carbon dioxide and water. Write a word equation for the reaction.

**D** Copper carbonate decomposes on heating to make copper oxide and carbon dioxide. Write a word equation for the reaction.

## Chemical reactions

A teacher heats a small piece of sodium. When the sodium catches fire, she places it in a flask of chlorine. The sodium continues to burn. White fumes are produced.

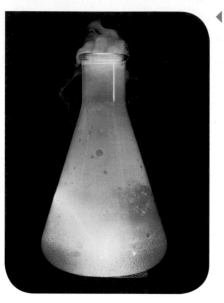

◀ Sodium reacts vigorously with chlorine

There has been a **chemical reaction**. Atoms of sodium and chlorine have joined together to make a new substance, sodium chloride.

There are millions of possible chemical reactions. In all of them, the atoms of the starting materials, or **reactants**, are rearranged to make new substances, or **products**. The properties of the products are different from those of the reactants.

Most chemical reactions are irreversible. Once a reaction has happened, it is difficult to get the starting materials back again.

## Word equations

The teacher uses a word equation to summarise the reaction of sodium and chlorine:

sodium + chlorine → sodium chloride

Word equations show the reactants and products of chemical reactions. But they do not tell us much else. To explain how the atoms are rearranged in a reaction, or to work out the amounts of substances that react together, you need a symbol equation.

## Formulae

Before you can write a symbol equation, you need to know the symbols or formulae of the reactants and products. For the reaction of sodium with chlorine:

- The symbol of the element sodium is Na.
- The symbol of the element chlorine is Cl. In chlorine gas the atoms are joined together in pairs to make chlorine molecules. So we represent chlorine gas by the formula $Cl_2$.
- The formula of sodium chloride is NaCl. This shows that the compound is made up of atoms of two elements. For every atom of sodium, there is one atom of chlorine. The **formula** tells us the number and type of atoms that are joined together in the compound.

Each compound has its own formula. The formula of carbon dioxide is $CO_2$. This shows that a molecule of the compound is made up of atoms of two elements, carbon and oxygen. For every one atom of carbon, there are two atoms of oxygen.

▲ A molecule of chlorine consists of two chlorine atoms strongly joined together

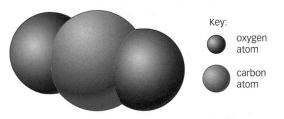

Key:
- oxygen atom
- carbon atom

▲ A molecule of carbon dioxide consists of one carbon atom and two oxygen atoms

E A molecule of nitrogen gas consists of two nitrogen atoms joined together. Write the formula of nitrogen gas.

F The compound lithium fluoride is made up of one lithium ion for every fluoride ion. Write the formula of lithium fluoride.

## Questions

1 Write a word equation for the reaction of potassium with fluorine to make potassium fluoride. ↓ E

2 Write the symbol of the element chlorine.

3 Write the formula of chlorine gas.

4 A molecule of sulfur trioxide consists of one atom of sulfur joined to three atoms of oxygen. Work out its formula. ↓ C

5 The formula of potassium manganate(VII) is $KMnO_4$. Explain what the formula tells us about this compound. ↓ A*

# 5: Chemical equations

▲ Burning titanium has started several fires in military jet plane engines

## Titanium fire

Titanium metal is important in the aerospace industry. Most of its properties make it an ideal material for jet engine parts. There's just one problem. At high temperatures, it catches fire. And moving parts in jet engines quickly reach high temperatures.

## Symbol equations

The word equation for the burning reaction of titanium is

titanium + oxygen → titanium dioxide

Atoms cannot be lost or made in chemical reactions. So if 48 g of titanium reacts with 32 g of oxygen, the mass of titanium dioxide formed will be 48 g + 32 g = 80 g.

A **balanced symbol equation** tells us more about the reaction by showing:

- how the atoms are rearranged
- the relative amounts of the substances that take part in the reaction.

Here's how to write a balanced symbol equation for the reaction of burning titanium:

- Write a word equation for the reaction. Put the correct symbol or formula under each reactant and product.

titanium + oxygen → titanium dioxide

$Ti + O_2 → TiO_2$

- Now balance the equation. The equation must show the same amounts of each substance on both sides of the arrow. Here, there are two atoms of oxygen and one atom of titanium on each side of the arrow. The equation is balanced.

> A List two things that a balanced symbol equation shows.
>
> B The balanced symbol equation for the burning reaction of carbon to make carbon dioxide is $C + O_2 → CO_2$. Explain what the equation tells us about the reaction.

## Laptop fire

It's 2006. Marv is using his laptop computer. Suddenly, it catches fire. Marv quickly extinguishes the flames, but the computer is ruined. What caused the fire?

It turns out that Marv wasn't alone. Many laptops caught fire at around the same time. Engineers realised that their lithium batteries were to blame. Sometimes, the batteries short-circuited, causing a spark. The spark set fire to tiny pieces of lithium metal floating in the battery's liquid. Then the fire spread.

Here's how to write a balanced symbol equation for the reaction:

- Write a word equation. Write a symbol or formula under each substance.

$$\text{lithium} + \text{oxygen} \rightarrow \text{lithium oxide}$$
$$\text{Li} + \text{O}_2 \rightarrow \text{Li}_2\text{O}$$

- Balance the amounts of oxygen. There are two atoms of oxygen on the left of the arrow and one on the right. Write a big number 2 to the left of the formula of lithium oxide.

$$\text{Li} + \text{O}_2 \rightarrow 2\text{Li}_2\text{O}$$

- The big 2 applies to each type of atom in the formula that follows it. Here, it means there are $(2 \times 2) = 4$ atoms of lithium and $(1 \times 2) = 2$ atoms of oxygen. The numbers of oxygen atoms are now balanced.

- Balance the amounts of lithium by writing a big 4 to the left of the symbol for lithium. There are now 4 atoms of lithium on each side. The equation is balanced.

$$4\text{Li} + \text{O}_2 \rightarrow 2\text{Li}_2\text{O}$$

## Questions

1  Write a word equation for the burning reaction of sulfur to make sulfur dioxide, $SO_2$.

2  Describe in words what the equation below tells us about the reaction of magnesium with oxygen.
$$2\text{Mg} + \text{O}_2 \rightarrow 2\text{MgO}$$

3  Sodium burns in oxygen to make sodium oxide, $Na_2O$. Write a balanced symbol equation for the reaction.

4  Write a balanced symbol equation to show that magnesium reacts with hydrochloric acid, HCl, to make hydrogen gas, $H_2$, and magnesium chloride, $MgCl_2$.

↓ E

↓ C

↓ A*

▲ Limestone buildings in Royal Crescent, Bath

▲ This limestone statue has been damaged by acid rain

## Rock of beauty

Every year, more than a quarter of a million people visit the city of Bath in southwest England. They are drawn to its Roman baths and its beautiful honey-coloured buildings. Many of these buildings were constructed more than 200 years ago. Most residents welcome the tourists – they bring in money and sustain jobs.

The houses and shops of the city are made from Bath stone. Bath stone is a type of **limestone**. Limestone was formed from the shells of creatures that lived in shallow seas between 440 and 70 million years ago.

Limestone is an important building material. It is attractive, durable, and strong. It can be cut into blocks. Some limestone is crushed into small lumps to make **aggregate**. Aggregate is used as a firm base beneath railway lines and roads.

Limestone is also a raw material for the production of other useful building materials, including cement, mortar, and glass.

## Acid attack

Unfortunately the surface of limestone can be damaged by chemical reactions with acid rain (see spread C1.15). This may make gaps between blocks in buildings.

The damage happens because limestone consists mainly of calcium carbonate, $CaCO_3$.

Calcium carbonate reacts with acids to make a salt, water and carbon dioxide. For example with sulfuric acid, an acid rain acid, the equation is:

$$\text{calcium carbonate} + \text{sulfuric acid} \rightarrow \text{calcium sulfate} + \text{carbon dioxide} + \text{water}$$

$$CaCO_3 + H_2SO_4 \rightarrow CaSO_4 + CO_2 + H_2O$$

Other carbonates react in similar ways with acids. For example:

$$\text{magnesium carbonate} + \text{hydrochloric acid} \rightarrow \text{magnesium chloride} + \text{carbon dioxide} + \text{water}$$

> A Describe three properties of limestone. Explain why these properties make limestone a good building material.

## Quarry queries

Companies get limestone from quarries – large holes in the ground. They use explosives to break up the rock before digging it out. Quarries bring benefits and problems.

### Benefits

Many quarries are in the countryside. Here, they provide jobs in places where work may be scarce. This helps local families and facilities such as shops and schools. These are **social impacts** of quarrying.

Products from quarries are valuable. Each year, British quarries produce materials worth several billion pounds. Although the UK imports some limestone, it exports even more. This contributes to the nation's economy. These are **economic impacts** of digging rock from the ground.

### Problems

Some quarries are in attractive areas of the countryside, where they may damage the tourist industry. Transporting rock from quarries to customers creates extra traffic, which may pass through small towns or villages.

Quarries also have **environmental impacts**. For example they take up land space, making the land unavailable for other uses such as farming and recreation.

> **B** Describe one social benefit and one social problem caused by quarrying.
>
> **C** Describe an economic impact of quarrying limestone.

▲ Limestone quarry

### Did you know...?

In the UK, we produce about 80 million tonnes of the rock every year. That's more than one tonne for every man, woman, and child.

### Key words

limestone, aggregate, social impact, economic impact, environmental impact

### Exam tip

✓ Always relate uses of materials to their properties.

### Questions

1  Calcium carbonate is the main mineral in limestone. Write down its formula and explain what it means.  ↓ E

2  Write a word equation to summarise the reaction of calcium carbonate.  ↓ C

3  Describe the social, economic, and environmental impacts of quarrying limestone. Explain which impacts are benefits and which are problems.  ↓ A*

▲ Heating limestone produces solid calcium oxide and carbon dioxide gas

## Making calcium oxide

Nadeem weighs a lump of limestone. He heats it for five minutes in a Bunsen flame. The limestone gets so hot that it glows. Nadeem lets the rock cool down. Then he weighs it again. The mass has decreased. Why?

On heating, calcium carbonate breaks down to make two new materials – calcium oxide and carbon dioxide gas. This is a **thermal decomposition** reaction.

calcium carbonate → calcium oxide + carbon dioxide

The balanced symbol equation for the reaction is:

$$CaCO_3(s) \rightarrow CaO(s) + CO_2(g)$$

The (s) shows that calcium carbonate and calcium oxide are both solids under the conditions of the reaction. Carbon dioxide is formed as a gas. The (s) and (g) are **state symbols**.

> **A** Name the products of the thermal decomposition reaction of calcium carbonate.
>
> **B** Give the states of each of the products of the reaction.

## Making calcium hydroxide

Nadeem goes back to his calcium oxide. He adds water to it, drop by drop. There is a chemical reaction. The product is calcium hydroxide. The reaction gives out so much heat it makes the water boil as Nadeem adds it.

calcium oxide + water → calcium hydroxide

$$CaO(s) + H_2O(l) \rightarrow Ca(OH)_2(s)$$

The (l) shows that the water is liquid.

> **C** Give the meanings of the symbols (s), (l), and (g) in symbol equations.

## Completing the lime cycle

Nadeem adds more water to the calcium hydroxide. Some of the calcium hydroxide dissolves. Nadeem filters the mixture and collects a colourless solution of calcium hydroxide. The solution is also called **limewater**.

Limewater is the test for carbon dioxide gas. Nadeem blows into the limewater through a straw. The limewater goes cloudy. The tiny pieces of solid that make the solution look cloudy are calcium carbonate. Nadeem has made the material he started with.

$$\text{calcium hydroxide} + \text{carbon dioxide} \rightarrow \text{calcium carbonate} + \text{water}$$

$$Ca(OH)_2(aq) + CO_2(g) \rightarrow CaCO_3(s) + H_2O(l)$$

The (aq) shows that the calcium hydroxide is dissolved in water.

Nadeem's series of reactions together make the **lime cycle**. The diagram below summarises this cycle.

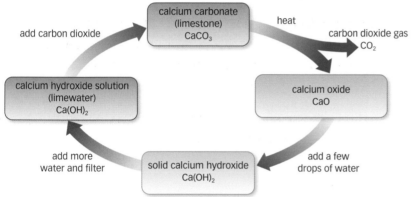

## Using products from the lime cycle

Calcium oxide and calcium hydroxide both form alkaline solutions with water. They have many uses:

- neutralising excess acidity in lakes and soils
- neutralising acidic waste gases produced by burning coal in power stations (see spread C1.15).

◀ Calcium oxide and calcium hydroxide neutralise acidic lakes

UXBRIDGE COLLEGE
LEARNING CENTRE

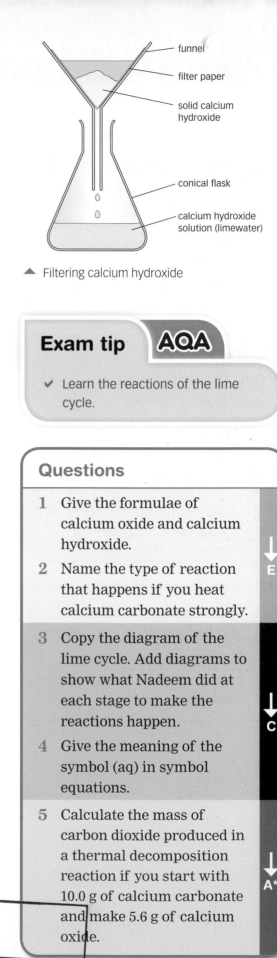

▲ Filtering calcium hydroxide

## Questions

1 Give the formulae of calcium oxide and calcium hydroxide.

2 Name the type of reaction that happens if you heat calcium carbonate strongly.  ↓ E

3 Copy the diagram of the lime cycle. Add diagrams to show what Nadeem did at each stage to make the reactions happen.  ↓ C

4 Give the meaning of the symbol (aq) in symbol equations.

5 Calculate the mass of carbon dioxide produced in a thermal decomposition reaction if you start with 10.0 g of calcium carbonate and make 5.6 g of calcium oxide.  ↓ A*

## Key words

cement, mortar, concrete, reinforced concrete

**A** List the raw materials for making cement.

**B** Suggest one reason for heating the kiln when making cement.

## Introduction

What do these structures have in common: the world's tallest building (the Burj Khalifa tower in Dubai), a brick house in the UK, and water pipes?

▲ Cement – a dry grey powder – was used in the construction of them all

## Cement

Companies make **cement** like this:

- Crush limestone rock into small pieces.
- Add powdered clay.
- Heat the mixture to 1450 °C in a rotating kiln.
- Add a little calcium sulfate powder.

During the heating stage, calcium carbonate in the limestone decomposes to make calcium oxide (quicklime) and carbon dioxide.

## Mortar

Bob is a bricklayer. He uses **mortar** to stick bricks together. He makes the mortar by mixing a thick paste from cement, sand, and water. Mortar sets overnight as substances in the mixture react with each other.

◀ Bricklayers use mortar to stick bricks together

## Concrete

Builders used 110 000 tonnes of **concrete** to make the foundations of the Burj Khalifa tower. They used an extra 39 000 tonnes of concrete to build the rest of the tower. The foundations and the tower are reinforced by huge steel bars.

Construction workers make concrete by mixing cement, sand, aggregate (small stones), and water. Like mortar, concrete sets as a result of chemical reactions within the mixture.

Concrete can form structures of many different shapes. It is strong under forces of compression (squashing), but weak if bent or stretched. It can be made much stronger by reinforcing it with steel. Because of these properties, concrete is the main material in millions of buildings. Bridges, mains water pipes, and paths are also made from concrete or **reinforced concrete**.

## Decomposing metal carbonates

Calcium carbonate decomposes when limestone is heated to make cement. The carbonates of sodium, magnesium, zinc, and copper also break down in thermal decomposition reactions.

For example, copper carbonate produces two products when you heat it strongly – copper oxide and carbon dioxide. Copper carbonate is green and copper oxide is black.

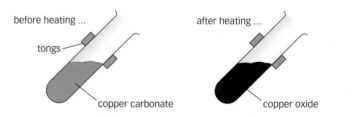

before heating ...   tongs   copper carbonate

after heating ...   copper oxide

The equation for the reaction is:

copper carbonate   →   copper oxide   +   carbon dioxide

$$CuCO_3(s) \quad \rightarrow \quad CuO(s) \quad + \quad CO_2(g)$$

Thermal decomposition reactions follow a pattern.
- Carbonates of metals low in the reactivity series (see spread C1.9), such as copper carbonate, need relatively small amounts of energy to break them down.
- The carbonates of metals high in the reactivity series, such as sodium carbonate, need a lot of energy to make them decompose.
- The carbonates of very reactive metals, such as potassium carbonate, do not decompose at Bunsen burner temperatures.

C  Describe two advantages of using concrete as a building material.

D  Explain why concrete structures have steel bars inside them.

E  Describe one change you would see if you heated copper carbonate strongly.

## Questions

1  Describe the appearance of cement.

2  List the raw materials used for making mortar and concrete.

3  What happens inside mortar and concrete to make them harden?

4  Write a word equation for the thermal decomposition reaction of sodium carbonate.

5  Describe the pattern shown by the thermal decomposition reactions of metal carbonates.

6  Write a balanced symbol equation for the thermal decomposition reaction of zinc carbonate, $ZnCO_3$.

↓ E

↓ C

↓ A*

## Learning objectives

After studying this topic, you should be able to:

✔ describe typical transition metal properties

✔ describe and explain how to extract gold and iron from the Earth's crust

## Metals for building

Imagine watching a football match in an aluminium-clad stadium, or enjoying art in Spain's titanium-covered museum. Or how about learning science in a copper-clad building in the Arctic? The properties of the metals used make them ideal for these buildings.

> **A** Suggest why the architects chose to cover the buildings in these metals.

▲ The copper-clad science centre in Svalbard, Norway

▲ The aluminium-clad Monterrey stadium in Mexico

▲ Spain's titanium-covered Guggenheim museum

## Metals in the middle

Titanium and copper are in the central block of the periodic table. Together, all the metals in this block are the **transition metals**. Like most metals, transition metals:

- have a shiny surface when freshly cut
- can be bent or hammered into different shapes without cracking
- are good conductors of heat and electricity.

## Did you know...?

Most gold jewellery is not pure gold. It is mixed with other metals to make it harder.

## Key words

**transition metal, unreactive, reactivity series, mineral, ore, blast furnace, reduction**

|  |  |  |  |  |  |  |  |  |  |  |  |  |  |  |  |  | 0 |
|---|---|---|---|---|---|---|---|---|---|---|---|---|---|---|---|---|---|
| 1 | 2 |  |  |  |  | H |  |  |  |  |  | 3 | 4 | 5 | 6 | 7 | He |
| Li | Be |  |  |  |  |  |  |  |  |  |  | B | C | N | O | F | Ne |
| Na | Mg |  |  |  |  |  |  |  |  |  |  | Al | Si | P | S | Cl | Ar |
| K | Ca | Sc | Ti | V | Cr | Mn | Fe | Co | Ni | Cu | Zn | Ga | Ge | As | Se | Br | Kr |
| Rb | Sr | Y | Zr | Nb | Mo | Tc | Ru | Rh | Pd | Ag | Cd | In | Sn | Sb | Te | I | Xe |
| Cs | Ba | La | Hf | Ta | W | Re | Os | Ir | Pt | Au | Hg | Tl | Pb | Bi | Po | At | Rn |
| Fr | Ra | Ac |  |  |  |  |  |  |  |  |  |  |  |  |  |  |  |

transition metals

# Where do metals come from?

## Gold

Why was gold one of the first metals to be discovered and used? Probably because it was easy to find – humans couldn't miss the grains and nuggets of the metal shining in the stream beds of ancient Egypt and the area that is now Iraq.

Gold is found as the metal itself because it is **unreactive**. It hardly ever joins to other elements to form compounds. It is near the bottom of the **reactivity series**.

## Iron

Iron – like most metals – is too reactive to exist on its own in the Earth's crust. It is joined to other elements in naturally occurring compounds called **minerals**. One important iron mineral is haematite – iron(III) oxide, $Fe_2O_3$. Minerals do not usually exist alone. They are mixed with sand or rock. Rocks that contain useful minerals are **ores**.

Iron is an important metal. Every year companies dig millions of tonnes of iron ore from the ground. They use chemical reactions to extract iron metal from the ore. Here's how:

- Put the iron ore (mainly iron(III) oxide) in a hot **blast furnace** with coke (carbon).
- Oxygen is removed from the iron(III) oxide in **reduction** reactions. The products are iron and carbon dioxide.

## Other metals

Other metals below carbon in the reactivity series are also extracted by heating their oxides with carbon. Carbon reduces metal oxides. For example:

$$\text{tin oxide} \quad + \quad \text{carbon} \quad \rightarrow \quad \text{tin} \quad + \quad \text{carbon dioxide}$$

### Questions

1 List three typical properties of transition metals.
2 Name two metals that are extracted from their minerals by heating with carbon.
3 What is a reduction reaction?
4 Write a word equation for the reduction of lead oxide by carbon to make lead and carbon dioxide.

most reactive

sodium
calcium
magnesium
aluminium
carbon
zinc
iron
tin
lead
copper
silver
gold
platinum

least reactive

▲ The reactivity series lists metals in order of their reactivity. Carbon is included to show its reactivity, even though it is not a metal.

## Exam tip · AQA

✔ Metals that are less reactive than carbon can be extracted from their oxides by reduction with carbon.

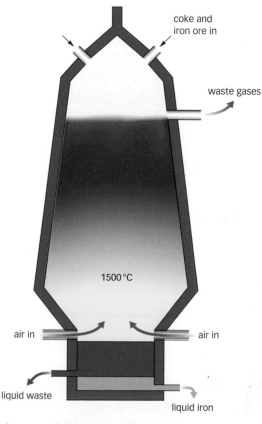

coke and iron ore in

waste gases

1500 °C

air in   air in

liquid waste

liquid iron

▲ Iron oxide is reduced in the blast furnace to make iron

## Learning objectives

After studying this topic, you should be able to:

- ✔ describe how atom arrangements in iron and steels are linked to their properties and uses
- ✔ explain the meaning of the word alloy

▲ A cast iron cannon

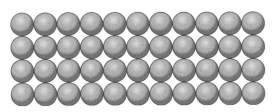

▲ The diagram shows one way of modelling the atom arrangement in pure iron

## Key words

cast iron, steel, alloy, low carbon steel, high carbon steel, corrosion, stainless steel

**A** Explain why pure iron is not useful.

**B** Explain why steel is an example of an alloy.

## Issues with iron

Iron from the blast furnace is not pure. It is a mixture of about 96% iron, 3% carbon, and other impurities. The impurities make blast furnace iron brittle – it breaks easily when you drop it. Blast furnace iron is not much use as it is.

### Cast iron

Re-melting blast furnace iron and adding scrap steel makes **cast iron**. Cast iron has a high strength in compression – you can press down on it with a great force and it will not break. Because of this property, cast iron was made into cannons and cooking pots.

### Pure iron

Removing the impurities from blast furnace iron makes pure iron. Pure iron has a regular arrangement of atoms.

The layers of atoms in pure iron slide over each other easily. This makes it easy to bend it into different shapes. Pure iron is also soft, which means it is not very useful.

So why do companies extract iron from its ores? How do they change its properties to make it useful?

## Steel – a vital alloy

**Steel** makes many things – from stunning structures to tiny components.

▲ The Millau viaduct is made from steel

▲ A steel screw

Steel is mainly iron. The iron is mixed with certain amounts of carbon, and sometimes other metals, to change its properties and make it more useful. There are many types of steel.

Steels are examples of **alloys**. An alloy is a mixture of a metal with one or more other elements. The physical properties of an alloy are different from the properties of the elements in it.

## Inside steel

The properties of an alloy depend on its atom arrangement. The atoms of carbon, iron, and other metals are of different sizes. In steel, the carbon and other metal atoms get between the iron atoms. They distort the regular pattern. The layers of iron atoms can no longer slide over each other easily. So steel has different properties from pure iron. Steel is harder and less bendy.

## Right for the job

People have been making steel for centuries. Bracelets that are 15 000 years old have been discovered in Tanzania. A thousand years ago, Chinese people made sharp steel swords.

### Carbon steels

Over the years, scientists experimented with different mixtures to make steels with perfect properties for particular purposes. They found that **low carbon steels**, containing less than 0.3% carbon, are easy to make into different shapes. Steel companies make low carbon steel sheets for car body panels and food cans.

**High carbon steels** have different properties. They are hard and strong. High carbon steels contain between 0.6% and 1.0% carbon.

### Stainless steel

Iron and steel go rusty when surface iron atoms react with water and oxygen from the air. The process is called **corrosion**. Corrosion weakens and damages iron and steel structures. It costs money and even lives.

**Stainless steels** do not go rusty. They are resistant to corrosion. They have this property because they contain chromium atoms.

Stainless steel makes cutlery, surgical instruments, and even the kitchen sink.

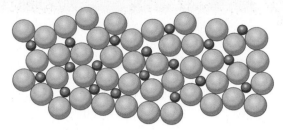

▲ A model of the structure of steel

▲ These surgical instruments are made from stainless steel

**Exam tip**

✓ Each type of steel has unique properties because of the arrangements of its atoms.

**Questions**

1. Use ideas about properties to explain why blast furnace iron is not useful. ↓ E

2. Summarise the properties and uses of low carbon steel, high carbon steel, and stainless steel. ↓ C

3. Use diagrams to explain how the atom arrangements of iron and steel explain their different properties. ↓ A*

## Learning objectives

After studying this topic, you should be able to:

✔ explain how the properties of copper make it suitable for its uses

✔ describe how copper is extracted from its ores and from waste

## Key words

open-cast mine, concentrate, smelting, electrolysis, low-grade ore, displacement, phytomining, bioleaching

**A** Give three uses of copper and explain how its properties make it suitable for each use.

An open-cast copper mine

## Copper crooks

A gang dresses up in high visibility jackets. They set up a line of traffic cones and rip out underground cables. Criminals are stealing as much copper telephone cable as they can carry away. Why? What makes copper so valuable?

## Pipes, pots, and pounds

Copper is valuable because its properties make it useful.

Chris is a plumber. He can choose either copper or plastic water piping – both are waterproof. Chris much prefers copper piping – he can bend it to the angles he needs and weld its joints to prevent leaks.

Sharon is an electrician. She uses many metres of plastic-covered copper wire every week, since copper is an excellent conductor of electricity.

Waqar is a cook. He uses copper-bottomed pans because copper conducts heat well.

Copper is also used to make coins. Five, ten, and twenty pence pieces are an alloy of copper with nickel. By itself, copper is too soft for coins. Mixing copper with nickel makes a harder alloy.

## Copper from the Earth's crust

Most copper in the Earth's crust is joined to other elements in minerals. The minerals are mixed with other substances in ores. Companies use several processes as they extract copper from its ores:

- They dig the ore from the ground at an **open-cast mine**.
- They **concentrate** the ore to separate copper minerals from the waste rock of the ore.

The companies then obtain pure copper by either or both of these two processes:

- They heat the concentrated ore in a furnace. Chemical reactions remove other elements from the copper. This is **smelting**. They then purify the copper by **electrolysis**.
- They make a solution of copper compounds from the ore and obtaining copper metal by electrolysis.

## Copper conundrum

Copper-rich ores are running out. So companies now extract copper from **low-grade ores** that contain little copper. This costs more than using copper-rich ores, but the high demand for copper means that they can still make a profit. But extracting copper from low-grade ores produces huge amounts of waste. The waste damages the environment.

These days, about half British copper is recycled, often by electrolysis. Copper can also be obtained from solutions of copper salts by adding scrap iron to the solutions. Chemical reactions result in the formation of copper metal. For example:

iron + copper sulfate → iron sulfate + copper

This is a **displacement** reaction.

Obtaining copper from solutions of its salts uses less energy and causes less pollution than extracting copper from its ores.

Scientists have also been researching new ways of extracting copper from low-grade ores without damaging the environment too much. The new techniques include:

- **Phytomining** – This involves planting certain plants on low-grade copper ores. The plants absorb copper compounds. Burning the plants produces ash that is rich in copper compounds.

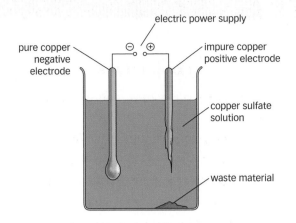

▲ The negative electrode is made of pure copper. Impure copper forms the positive electrode. During electrolysis, positive copper ions move to the negative electrode. Waste material falls to the bottom.

▲ The 'copper flower' plant absorbs large amounts of copper

- **Bioleaching** – Some bacteria can obtain their nutrients and energy from copper compounds in copper ores. The bacteria produce solutions of copper compounds. Chemical reactions or electrolysis extract copper metal from these solutions. The process is very slow.

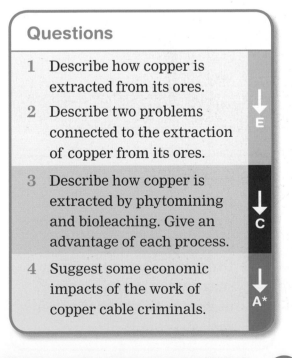

### Questions

1 Describe how copper is extracted from its ores.

2 Describe two problems connected to the extraction of copper from its ores.

3 Describe how copper is extracted by phytomining and bioleaching. Give an advantage of each process.

4 Suggest some economic impacts of the work of copper cable criminals.

## Learning objectives

After studying this topic, you should be able to:

✔ give reasons for the uses of aluminium, titanium, and their alloys

✔ explain why the metals are expensive to extract

✔ explain the benefits of recycling metals

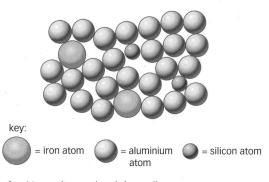

key:

 = iron atom    = aluminium atom    = silicon atom

▲ Atoms in an aluminium alloy

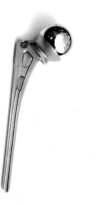

▲ An artificial hip made from titanium

**A** Explain which properties of titanium make its alloys suitable for making aeroplane parts.

**B** Use ideas about properties to suggest why aluminium is used to make overhead power cables.

## Wheels and wings

Jason's car has alloy wheels. They were much more expensive than normal steel wheels. So why did he buy them?

First, Jason thinks shiny alloy wheels look good. He likes their exotic shapes and styles, too.

Second, the wheel alloy is mainly aluminium. Aluminium has a low density – it is light for its size. So alloy wheels are lighter than steel wheels of the same size.

Aluminium has a thin layer of aluminium oxide on its surface. This layer stops oxygen and water molecules reacting with the aluminium atoms underneath. So aluminium does not corrode.

Aluminium on its own would not make good wheels – the metal is too soft. But an alloy of 93% aluminium mixed with silicon and iron is perfect for the job. The silicon and iron atoms disrupt the pattern of atoms in pure aluminium. Aluminium atoms slide over each other less easily in the alloy, so the alloy is harder than the pure metal. Aluminium alloys are also much stronger than pure aluminium.

The properties of aluminium mean that the metal and its alloys are also used in:

- aeroplanes
- overhead power cables
- cooking foil
- drinks cans.

## Using titanium

Titanium is a transition metal. It has typical transition metal properties. And, like aluminium, it has a low density and resists corrosion.

These properties mean that titanium alloys are useful for making aeroplanes, even though titanium catches fire more easily than some metals (see spread C1.5).

Titanium is also used to make artificial hips and bone pins, and for oil rigs at sea.

# Extracting titanium and aluminium

Most of the aluminium we use came originally from aluminium oxide in bauxite ore. Most titanium exists naturally as titanium oxide. There are limited amounts of the ores of both metals – they won't last forever.

Aluminium and titanium are expensive. This is because they are not easy to extract. You can't extract aluminium or titanium from their ores by heating with carbon. Here's why:

- Aluminium is above carbon in the reactivity series. Its atoms are joined very strongly to oxygen atoms in bauxite.
- If you heat titanium oxide with carbon you make titanium carbide. This makes the metal brittle.

So aluminium is extracted by electrolysis. The process uses a lot of electrical energy, which is expensive.

Titanium is extracted from its ores in a multi-step process. Again, much energy is needed, so the process is expensive.

> C  Explain why titanium and aluminium are expensive.

## Reduce, reuse, recycle

Many people recycle aluminium cans. Creating new cans from aluminium uses only about 10% of the energy needed to make cans from newly extracted bauxite. Less pollution is created, too.

▲ A bauxite mine

▲ Titanium ore

**Did you know...?**

Titanium artificial joints are so resistant to corrosion that they can stay in place for 20 years.

**Exam tip**

- ✔ Aluminium and titanium are useful because they have low densities and they do not corrode. They are expensive because they are extracted from their ores in multi-step processes which need lots of energy.

## Questions

1  Explain which properties of titanium make it suitable for artificial hips.  ↓ E

2  Explain why aluminium cannot be extracted from its ore by heating the ore with carbon.  ↓ C

3  List and explain two advantages of recycling aluminium.

4  Use ideas about particles and properties to explain why aluminium alloys are used in aeroplanes, not the pure metal.  ↓ A*

> **A** Name three fossil fuels.
>
> **B** Explain why oil is a finite resource.

## Limited supplies

Have you washed, got dressed, or travelled today? If so, you have probably used products made from **crude oil**, like those pictured here. Crude oil is a vital fuel and raw material. Oil companies extract millions of tonnes of it every day.

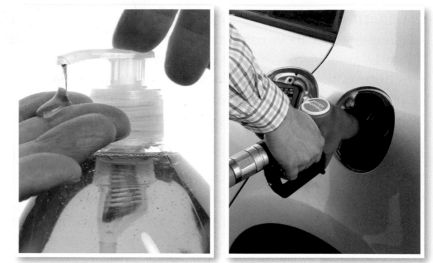

Crude oil is a **fossil fuel**. It was formed from the decay of buried dead sea creatures. The process took millions of years. We use up oil faster than new oil forms. So oil is **non-renewable**.

Oil wasn't made just anywhere. The conditions had to be exactly right. So today's reserves are limited, and will run out one day. Oil is a **finite** resource.

Coal and natural gas are fossil fuels too – it took millions of years to make them. Coal was formed from trees that were buried under swamps. Natural gas, like oil, was formed from dead sea creatures.

## What's in crude oil?

Crude oil contains many different **hydrocarbons**. Hydrocarbons are compounds made up of hydrogen and carbon only.

Crude oil is a **mixture**. This means its different hydrocarbon compounds are not chemically joined together. Each of the hydrocarbons has its own properties. Being part of a mixture does not affect these properties.

You can separate mixtures by physical means:
- Filtration separates a solid from a liquid.
- Distillation separates liquids with different boiling points.

> **C** Explain the meaning of the word hydrocarbon.
>
> **D** Give two characteristics of a mixture.

## Fractions

Crude oil is not much use as it is. But separate it into **fractions** and you get valuable fuels and raw materials. A fraction is a mixture of hydrocarbons with similar numbers of carbon atoms and similar boiling points.

## Separating oil fractions

Oil companies use the property of boiling point to separate crude oil into useful fractions by **fractional distillation**. The process is continuous – it carries on all the time.

Fractional distillation involves heating crude oil to about 450 °C. Its compounds **evaporate** to become gases. The gases enter the bottom of a **fractionating column**. The column has a temperature gradient. It is hot at the bottom and cooler at the top.

The gases move up the column. As they move up, they cool down. Different fractions **condense** to form liquids again at different levels.
- Compounds with the highest boiling points condense at the bottom of the column, and leave as liquids.
- Lower boiling point compounds condense higher up, where it is cooler, and leave as liquids.
- The lowest boiling point compounds leave from the top, as gases.

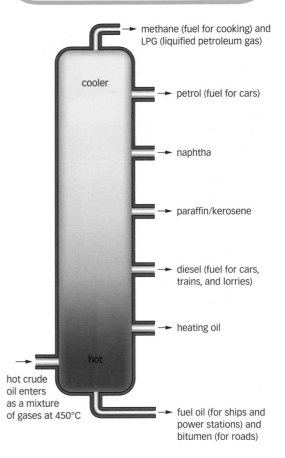

methane (fuel for cooking) and LPG (liquified petroleum gas)

cooler

petrol (fuel for cars)

naphtha

paraffin/kerosene

diesel (fuel for cars, trains, and lorries)

heating oil

hot

hot crude oil enters as a mixture of gases at 450°C

fuel oil (for ships and power stations) and bitumen (for roads)

▲ Fractionating column

## Questions

1. Give two characteristics of fossil fuels.
2. What is a crude oil fraction? Name three fractions, and give their uses.
3. Describe how crude oil is separated into fractions by fractional distillation. Use the words boil, evaporate, and condense in your answer.

## Exam tip    AQA

- ✔ Remember – high boiling temperature fractions are removed at the bottom of a fractionating column; lower boiling temperature fractions come out higher up. You do not need to remember the names of specific oil fractions.

## Learning objectives

After studying this topic, you should be able to:

✔ use molecular and displayed formulae to represent alkanes

✔ know the formulae of the first three alkanes

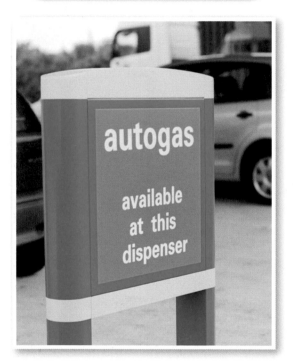

▲ Autogas is a form of LPG

## Exam tip  AQA

✔ Remember the patterns in properties of alkanes to help you answer questions about their suitability for use as fuels.

✔ Learn the formulae of methane, ethane, propane, and butane. You do not need to learn the formulae or names of other alkanes.

## Inside oil fractions

Some cars are fuelled by LPG, liquefied petroleum gas. Their exhaust emissions are cleaner than those from diesel cars.

Liquefied petroleum gas is obtained from crude oil. It is a mixture of two hydrocarbons with similar boiling points, propane and butane.

A propane molecule consists of three carbon atoms joined to eight hydrogen atoms. Its **molecular formula** is $C_3H_8$. You can also represent propane by its **displayed formula**. This shows how the atoms are arranged in its molecules. Each letter represents an atom. Each line represents a covalent bond between two atoms.

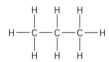

Butane molecules are made up of four carbon atoms joined to ten hydrogen atoms. The molecular formula of butane is $C_4H_{10}$. Its displayed formula is

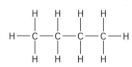

Propane and butane are **alkanes**. Alkanes are **saturated** hydrocarbons. This means their carbon atoms are joined together by single covalent bonds. Most other hydrocarbons in crude oil are alkanes, too.

- Methane is the main compound in natural gas. Its formula is $CH_4$.
- Ethane is also a gas. Its formula is $C_2H_6$.
- Hexane is part of the petrol fraction. Its formula is $C_6H_{14}$.

> **A** Draw displayed formulae for methane and ethane.
>
> **B** Give the numbers of carbon and hydrogen atoms in a molecule of hexane.

All alkanes have the same **general formula**, $C_nH_{2n+2}$. This means that the number of hydrogen atoms in an alkane is twice the number of carbon atoms plus two.

# Size matters

Th combustion of hydrocarbon fuels releases energy. The uses of alkanes as fuels depend on their properties. And their properties depend on the sizes of their molecules.

## Viscosity

Hexadecane molecules are long. They get tangled up. So hexadecane is difficult to pour and does not flow easily. It is a **viscous** liquid at room temperature. Pentane is also liquid at room temperature. Its smaller molecules make it runnier – or less viscous – than hexadecane. There is a pattern in the viscosity of liquid alkanes – the longer the hydrocarbon chain, the more viscous the liquid.

## Melting and boiling points

Molecule size also influences boiling points. Again, there is a pattern – the smaller the molecule, the lower the boiling point.

| Number of carbon atoms in hydrocarbon chain | State at room temperature |
|---|---|
| 1 to 4 (smaller molecule) | gas |
| 5 to 16 | liquid |
| 17 or more (larger molecule) | solid |

## Igniting alkanes

Alkanes with small molecules catch fire more easily than those with bigger molecules.

▲ Methane is a good fuel for cooking because it ignites easily

▲ Longer chain alkanes are more viscous because their molecules get more tangled

C Describe the properties of methane that make it a good fuel for cooking.

D Explain why the properties of hexadecane make it a suitable vehicle fuel when mixed with other hydrocarbons.

## Questions

1 Give the numbers of carbon and hydrogen atoms in a molecule of pentane, $C_5H_{12}$.

2 Draw a displayed formula for pentane.

3 Use the general formula for alkanes to work out the formula of an alkane with 15 carbon atoms.

4 Explain why an alkane with molecule formula $C_{18}H_{38}$ is not an ideal cooking fuel.

↓ E
↓ C
↓ A*

## Learning objectives

After studying this topic, you should be able to:

✔ identify hydrocarbon combustion products

✔ explain their impacts on the environment

▲ Fire dancers burn kerosene on their firesticks

▲ Didcot power station burns coal and gas

▲ These trees have been damaged by acid rain

## Dancing with fire

Aaden is a fire dancer. Backstage, he dips the wick of his firesticks into kerosene and sets them alight. Then he leaps onto the stage and thrills his audience with spectacular displays of twirling fire and hot dance moves.

Kerosene is a mixture of hydrocarbons. When it burns outside, there is plenty of oxygen around. Its combustion makes mainly carbon dioxide gas and water vapour. For example:

$$\text{undecane} + \text{oxygen} \rightarrow \text{carbon dioxide} + \text{water}$$
$$C_{11}H_{24} + 17O_2 \rightarrow 11CO_2 + 12H_2O$$

This is an example of **complete combustion**.

## Greenhouse gas

Carbon dioxide is a greenhouse gas. Find out more about its impacts on the environment on spread C1.16.

## Killer rain

Didcot power station burns coal and gas to generate electricity. Coal contains sulfur impurities. So when coal burns it produces **sulfur dioxide** gas as well as carbon dioxide and water. If sulfur dioxide goes into the air, it dissolves in water in clouds. This makes **acid rain** – rain which is more acidic than normal.

Oxides of nitrogen such as nitrogen dioxide, $NO_2$, are formed when hydrocarbon fuels burn at high temperatures in car engines. These also dissolve in water to make acid rain.

Acid rain has many environmental impacts:

- It makes lakes more acidic. Some species of water animals and plants cannot survive if the water is too acidic.
- Acid rain damages trees. It dissolves soil nutrients and washes them away before tree roots can take them in. It also damages the protective waxy coating of leaves. The leaves can no longer produce enough food for the tree.
- Acid rain damages limestone buildings by reacting with the calcium carbonate of the limestone (see spread C1.6).

There are signs that the UK is beating acid rain. Scientists have discovered how to remove sulfur impurities from fuels such as diesel.

It is less easy to remove sulfur impurities from coal. So some power stations remove sulfur dioxide from their waste gases by adding limestone powder to the gases. There is a chemical reaction. Calcium carbonate in limestone reacts with sulfur dioxide to make calcium sulfate. Calcium sulfate is a useful product. It is used to make plasterboard for houses.

> A  Name the gases that cause acid rain.
> B  Describe and explain three impacts of acid rain.
> C  Describe two ways of reducing acid rain.

## Particulate problem

This car runs on diesel fuel. Burning diesel produces tiny pieces of solid, or **particulates**, as well as carbon dioxide and water. Particulates may be made up of soot, a form of carbon, and unburned fuels.

Particulates may cause **global dimming**. In the atmosphere, particulates reflect sunlight back into space. As more particulates enter the atmosphere, less sunlight reaches the Earth's surface.

## Silent killer

Edith heats her house with a gas boiler. She hasn't had it checked recently. One day, her neighbour finds her confused, drowsy, and barely able to breathe. Edith has **carbon monoxide** poisoning.

The killer gas was produced because too little air was reaching the flame of the boiler. **Incomplete combustion**, or **partial combustion**, happened, so carbon monoxide, CO, was formed as well as carbon dioxide.

### Key words

complete combustion, sulfur dioxide, acid rain, particulates, global dimming, carbon monoxide, incomplete combustion, partial combustion

### Exam tip

✔ Remember – sulfur dioxide and oxides of nitrogen cause acid rain, particulates cause global dimming, and carbon monoxide is poisonous.

> D  Describe an environmental impact of particulates.

### Questions

1  Give two symptoms of carbon monoxide poisoning. ↓E

2  Explain the difference between complete and incomplete combustion. ↓C

3  Make a table to summarise the environmental impacts of three products of combustion of hydrocarbons. ↓A*

## Learning objectives

After studying this topic, you should be able to:

✔ describe and explain the impacts of the greenhouse gas carbon dioxide

## Key words

greenhouse gas, global warming, climate change, alternative fuel, carbon neutral

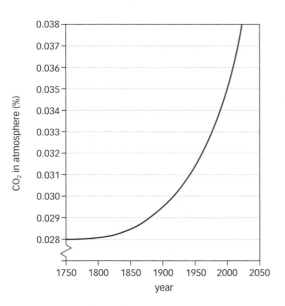

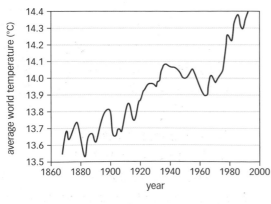

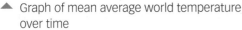
▲ Graph of carbon dioxide concentration over time

▲ Graph of mean average world temperature over time

## Greenhouse gas

Burning fossil fuels produce carbon dioxide. Carbon dioxide is a **greenhouse gas**. Its presence in the atmosphere helps keep Earth warm enough for life.

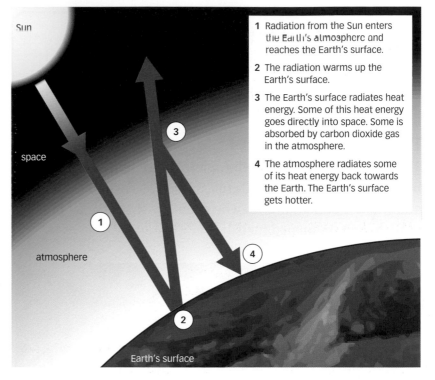

1 Radiation from the Sun enters the Earth's atmosphere and reaches the Earth's surface.

2 The radiation warms up the Earth's surface.

3 The Earth's surface radiates heat energy. Some of this heat energy goes directly into space. Some is absorbed by carbon dioxide gas in the atmosphere.

4 The atmosphere radiates some of its heat energy back towards the Earth. The Earth's surface gets hotter.

▲ Global warming

## Global warming – the cause

Over time, the average air temperature of the Earth has increased. This is **global warming**. At the same time, the percentage of carbon dioxide in the atmosphere has also increased. The graphs show these changes.

In the 1950s, scientists wondered if extra carbon dioxide might be causing the temperature increase. They began collecting evidence. Since the 1980s, hundreds of scientists have researched global warming. There is now a huge body of evidence supporting the theory that extra carbon dioxide from human activities causes global warming.

A Describe how greenhouse gases keep the Earth warm.

B Describe and explain what the graphs on the left show.

## Global warming – the impacts

The impacts of global warming include:

- **Climate change** – the Earth's weather patterns are changing. Some people are already dealing with more extreme weather events. Some areas will have more droughts. Others will have more flooding. Climate change puts some plant and animal species at risk of extinction. Other species – such as mosquitoes that carry the deadly disease of malaria – may spread over wider areas.
- **Melting ice caps** – higher average temperatures are making polar ice caps melt. Sea levels are rising and coastal areas – including big cities – are at risk of flooding.

▲ Climate change causes flooding

> **C** Describe and explain two impacts of global warming.

## Global warming – some solutions

### Using less

Governments, councils, and environment groups are urging people to consume less and use less fuel.

### Alternative fuels

Scientists are playing their part in trying to prevent climate catastrophe. They have developed cars fuelled by **alternative fuels**, such as hydrogen and ethanol.

Hydrogen cars produce one exhaust product – water vapour. But there are problems. Hydrogen fuel must be manufactured – either from methane or by using electricity to break up water molecules. Both these processes produce carbon dioxide gas. Storing and transporting hydrogen fuel – which is an explosive gas – are difficult and expensive.

In Brazil, many cars are fuelled by ethanol. The ethanol is made from crops such as sugar cane. Some people say that fuel produced from renewable crops is **carbon neutral**. Growing sugar cane takes carbon dioxide from the atmosphere. Burning ethanol returns a similar amount of carbon dioxide to the atmosphere. But the balance is not perfect. Energy is needed to make fertilisers and to manufacture ethanol. Both processes produce carbon dioxide gas. Some people think that it is unethical to use land to grow fuel instead of food.

> **Exam tip** **AQA**
>
> ✔ Make sure you know the difference between the terms 'global warming', 'greenhouse gas', and 'climate change'.

▲ Ethanol fuel can be produced from sugar cane

### Questions

1 Name a fuel produced from sugar cane. ↓E

2 Some people say that ethanol produced from sugar cane is carbon neutral. Explain why.

3 Draw and annotate a big diagram to explain the causes and effects of global warming. ↓C

4 Describe and explain the problems associated with producing and using hydrogen and ethanol fuels. ↓A*

**A** Explain why plant oils make good fuels.

**B** Use the data to identify which of the three plant oils transfers the most energy on burning.

**C** Suggest why plant oils are used for fuel even though they transfer less energy than diesel.

▲ Oil palms produce a useful fuel

## Plant oil fuels

Brendan has a diesel car. But he no longer buys expensive diesel fuel. Instead, he produces cheap fuel in his garage – from used cooking oil from the chip shop down the road. Brendan had to convert his car engine to prevent damage from the cooking oil fuel.

◀ Biofuel for cars

**Plant oils** make good fuels for cars, buses, and trains because they transfer large amounts of energy when they burn. The table compares plant oils to diesel.

| Fuel | Energy transferred on burning 1 kg of the fuel (kJ, approximate values) |
|---|---|
| diesel | 45 000 |
| sunflower oil | 38 000 |
| peanut oil | 40 000 |
| rapeseed oil | 37 000 |

Plant oils are not the only type of **biofuels** suitable for use in vehicles. Two other types are:

• ethanol produced from crops such as sugar cane (see spread C1.16)
• biodiesel produced by chemically reacting plant oils or animal fats with an alcohol.

## Palm oil palaver

In 2009, a company asked for permission to build a biofuel power station in Bristol. The power station would burn oil from oil palm trees as well as other plant oils. The burning reaction would heat water to make steam. The steam would be used to turn turbines. The turbines would generate enough electricity for 25 000 homes.

Many people expressed their opinions about the power station.

The power station will employ 30 local people. It will generate electricity from renewable sources.

It is nonsense to say that palm oil fuel is a leading cause for rainforest destruction.

In countries like Indonesia, tropical rainforests are destroyed to make space to grow oil palms. This damages biodiversity and has decimated orang-utan and elephant populations.

Burning plant oils results in emissions of oxides of nitrogen gases. These worsen asthma and other lung diseases.

One in six people in the world do not have enough to eat. It is immoral to use land to grow plants for fuel and not for food.

Oil palms remove carbon dioxide from the atmosphere as they grow. Similar amounts of carbon dioxide are returned to the atmosphere by burning palm oil.

A palm oil company evicted us from our land. We have nowhere to go.

We have collected data about carbon dioxide emissions. It shows that burning rainforests to grow oil palms emits thousands of times as much carbon dioxide as can be prevented by using palm oil.

In 2010, Bristol City Council decided not to give the go-ahead for the biofuel power station.

**D** Identify two benefits of using plant oils to generate electricity.

## Key words

**plant oils, biofuel**

## Exam tip — AQA

✓ Plant oil fuels are renewable. However, they produce carbon dioxide on burning. Growing plants for oil may lead to environmental destruction.

## Questions

1 Identify two ethical and two environmental arguments against using plant oils to generate electricity. ↓ E

2 Explain two drawbacks of using palm oil as a fuel.

3 Suggest two reasons for Bristol City Council not allowing the power station to be built. ↓ C

4 Write a paragraph to describe and explain the benefits, drawbacks, and risks of using plant oils as fuels. ↓ A*

# Course catch-up

## Revision checklist

- Each element in the periodic table is built up of one sort of atom.
- Atoms have a nucleus containing protons and neutrons surrounded by electrons within energy levels (shells).
- Elements in the same group of the periodic table have the same number of electrons in the highest energy level (outer shell). They react in similar ways.
- Balanced equations describe chemical reactions.
- Limestone (calcium carbonate) is a sedimentary rock and is quarried from the ground for use as a building material.
- Limestone is heated with clay to form cement. Cement, water, and sand form concrete.
- Metal carbonates are decomposed by heat to form metal oxides + carbon dioxide.
- Calcium oxide reacts with water to form calcium hydroxide. This reacts with carbon dioxide to reform calcium carbonate.
- The elements in the middle block of the periodic table are transition metals.
- Pure iron is converted into an alloy (steel) by mixing with other elements. This alters the arrangement of atoms.
- Metals are extracted from metal ores by electrolysis, heating, or reaction with carbon (reduction).
- Aluminium and titanium have low density and do not corrode. They are extracted using expensive electrolysis.
- Recycling metals saves resources, energy, and waste.
- Crude oil is a non-renewable resource. It is a mixture of hydrocarbons (separated by fractional distillation).
- Alkanes are saturated hydrocarbons which contain only single bonds and have general formula $C_nH_{2n+2}$.
- Burning hydrocarbon fuels releases carbon dioxide (greenhouse effect), carbon monoxide (toxic), sulfur oxides and nitrogen oxides (acid rain), particulates (global dimming).
- Biofuels (biodiesel, bioethanol) are produced from plant material. Producing biofuels releases less $CO_2$ overall and uses renewable resources. It may reduce the amount of food crops the world can grow.

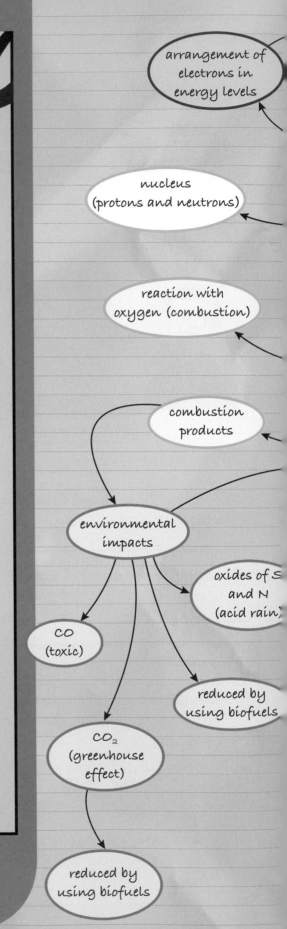

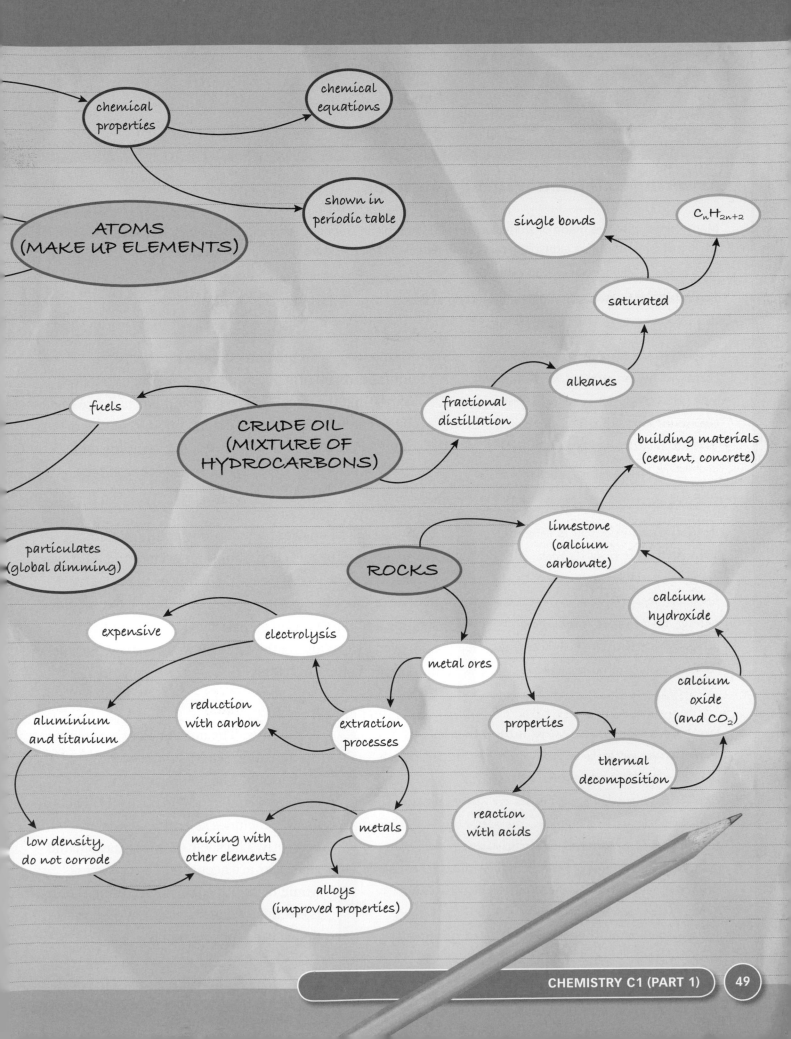

chemical properties

chemical equations

ATOMS (MAKE UP ELEMENTS)

shown in periodic table

single bonds

$C_nH_{2n+2}$

saturated

alkanes

fractional distillation

fuels

CRUDE OIL (MIXTURE OF HYDROCARBONS)

building materials (cement, concrete)

limestone (calcium carbonate)

particulates (global dimming)

ROCKS

calcium hydroxide

expensive

electrolysis

metal ores

calcium oxide (and $CO_2$)

aluminium and titanium

reduction with carbon

extraction processes

properties

thermal decomposition

low density, do not corrode

mixing with other elements

metals

reaction with acids

alloys (improved properties)

# Answering Extended Writing questions

Many drinks cans are made of aluminium. In 2008, people in Germany recycled 96% of their aluminium cans, while just 51% of British aluminium cans were recycled. The rest were buried in landfill sites.

Give reasons why Britain should increase the percentage of aluminium cans it recycles.

**The quality of written communication will be assessed in your answer to this question.**

---

**G – E**

If you resycle alyouminium cans there will be less climate warming. If you throw away cans on the street, they mite hurt squirrels and birds. Britain should resycle as much as Germany.

**Examiner:** The candidate knows that recycling aluminium results in less global warming than extracting the metal from its ore, but has not explained why. There is no scientific detail, and the candidate has muddled two scientific terms (global warming and climate change). There are several spelling errors.

---

**D – C**

Recycling aluminium is better than getting it from its rock because recycling makes less polution! Recycling also needs less energy, and makes less carbon diokside! Getting aluminium from its rock causes climate change!

**Examiner:** The answer includes four points explaining the advantages of recycling aluminium. The candidate uses several scientific words correctly, but uses the word 'rock' instead of 'ore'. The answer is well structured, but includes spelling and punctuation errors. The answer would be improved by explaining the links between energy, carbon dioxide, and climate change.

---

**B – A\***

Recycling more aluminium means that less aluminium needs to be produced from its ore, which is good since the ore is a finite resource. Extracting aluminium from its ore makes dangerous pollution. Recycling aluminium does not. Electricity is needed to get aluminium from its ore. Generating electricity makes carbon dioxide, which causes global warming. Recycling aluminium needs much less energy. Aluminium cans that are not recycled go to ugly landfill sites.

**Examiner:** The answer clearly explains the scientific reasons for recycling aluminium. The points are well explained and in a sensible order. The answer includes scientific words, used correctly. The spelling, punctuation, and grammar are accurate. The answer would be improved by mentioning that it is electricity generated from fossil fuels that causes carbon dioxide.

# Exam-style questions

**1** Choose metals from this list to answer the questions about the ways in which metals are extracted from their ores:

gold        iron        aluminium

**A01** **a** Which metal is extracted from its ore in the blast furnace?

**A01** **b** Which metal is extracted by electrolysis?

**A01** **c** Which metal is so unreactive that it is found naturally in the Earth's crust?

**2** Steel is an alloy made up of iron with added carbon. Adding carbon affects the properties of steel. This graph shows the results of an experiment done to find out how the strength of steel is affected by the amount of carbon added.

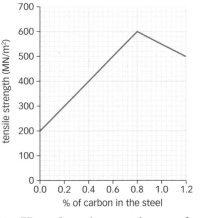

**A03** **a** How does increasing carbon percentage in the alloy affect its strength?

**A03** **b** Imagine that steel with a tensile strength of 500 MN/m² needs to be produced. Give two possible percentages of carbon which will produce a suitable steel.

**A01** **c** Give one other property of metals, apart from strength, which can be improved by using an alloy.

**3** A sodium atom has 11 electrons. It also contains protons and neutrons.

**A02** **a** Draw a labelled diagram to show how these particles are arranged in an atom of sodium.

**A01** **b** Give an example of a chemical property which is similar for all Group 1 elements.

## Extended Writing

**4** Limestone is a rock found in some
**A01** parts of the UK. Large quarries are dug to get the limestone out of the ground. Why is limestone an important resource, and why do some people object to limestone quarries?

**5** Metal carbonates, such as limestone,
**A02** decompose into carbon dioxide and metal oxides when heated. Some carbonates decompose at a higher temperature. Describe an experiment to find out which of two carbonates decomposes at the highest temperature.

**6** Biodiesel is an example of a biofuel. It
**A02** can be produced from plant oils such as
**A03** palms or rapeseed. Scientists debate whether the use of biofuels is a benefit to the environment. Discuss the arguments for and against the use of biofuels.

D–C

G–E

B–A*

G–E

D–C

B–A*

D–C

# C1 Part 2

# Polymers, plant oils, the Earth, and its atmosphere

## Why study this unit?

Every year, volcanoes, and earthquakes cause death and destruction. But what causes these disasters? Can scientists predict them and so save lives?

In this unit you will discover how gigantic pieces of the Earth's crust grind against each other to cause earthquakes, and find out how scientists monitor volcanoes to predict eruptions.

As well as providing fuel, crude oil provides most of the huge variety of plastics we use every day. How are these plastics made? How can we minimise their impact on the environment?

You will learn how chemicals from crude oil make plastics, and about their renewable replacements. You will also consider how burning fossil fuels changes our atmosphere, leading to acid rain and global warming.

## You should remember

1  In chemical reactions, atoms are rearranged to make new substances.

2  There are patterns in the chemical reactions of substances.

3  The way we use materials depends on their properties.

4  There are patterns in the properties of materials.

5  Humans extract many useful materials from plants.

6  The Earth consists of three layers – the crust, mantle, and core.

7  Earthquakes and volcanoes occur mainly in certain regions of the world.

8  The Earth's atmosphere is a mixture of gases.

The 1816 volcanic eruption of Mount Tambora in Indonesia killed up to 92 000 people, including 10 000 from the explosion and ash fall, and 82 000 from other causes. The year 1816 became known as 'the year without a summer', because volcanic ash in the atmosphere lowered global temperatures by an average of between 0.4 °C and 0.7 °C. Lower temperatures led to failed harvests and severe food shortages in many parts of the world.

This photo shows the July 1980 eruption of Mount St Helens, Washington.

## Learning objectives

After studying this topic, you should be able to:

✔ explain how cracking makes useful products

✔ identify the products of cracking reactions

✔ identify an economic benefit of cracking reactions

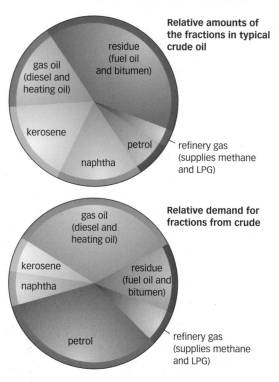

Relative amounts of the fractions in typical crude oil

Relative demand for fractions from crude

▲ These charts compare the supply and demand for crude oil fractions

▲ A catalytic cracking tower at an oil refinery

## Petrol, plastics, and propanone

What do the things in the pictures have in common? Petrol, polythene bottles, and propanone for nail varnish remover are all made from chemicals obtained from crude oil. Thousands of other important chemicals come from crude oil too.

## Cracking

The fractional distillation of crude oil makes useful products, including petrol. But the supply of some fractions does not match demand (the amounts people need).

Companies use **cracking** to break down bigger hydrocarbon molecules to smaller, more useful molecules.

Cracking involves heating an oil fraction to a high temperature. The hydrocarbons of the fraction vaporise. Then the vapour passes over a hot **catalyst**. Alkane molecules in the vapour break down to form smaller molecules in thermal decomposition reactions. A catalyst is a substance that speeds up the reaction. It is not used up in the reaction itself. Hydrocarbons can also be cracked by mixing their vapours with steam and heating to a very high temperature.

Octane has the formula $C_8H_{18}$. It is an alkane in the naphtha fraction. In cracking, each octane molecule breaks down to make two smaller molecules, such as hexane ($C_6H_{14}$) and ethene ($C_2H_4$). The equation for this cracking reaction is

$$C_8H_{18} \rightarrow C_6H_{14} + C_2H_4$$

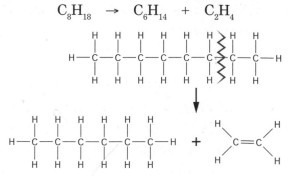

Hexane is an alkane. It is a useful fuel. Hexane is added to the petrol fraction, so the company now has more petrol to sell.

## Inside ethene

Ethene is a hydrocarbon. It is made up of atoms of carbon and hydrogen only.

You can represent ethene by its molecular formula, $C_2H_4$. Its displayed formula shows how its atoms are joined together.

The double line between the two carbon atoms represents a double covalent bond. Double bonds are stronger than single bonds. Because ethene has a double bond it is an **unsaturated hydrocarbon**. The double bond makes ethene more reactive than ethane.

You can detect compounds with double bonds by testing with a solution of bromine, or **bromine water**. Orange bromine water becomes colourless when it reacts with ethene and other unsaturated compounds.

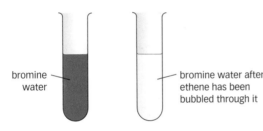

bromine water

bromine water after ethene has been bubbled through it

## The alkene family

Ethene belongs to a family of hydrocarbons called the **alkenes**. All alkenes are unsaturated. Another alkene is propene, $C_3H_6$

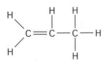

Alkenes have the general formula $C_nH_{2n}$. This shows that the number of hydrogen atoms in an alkene molecule is double the number of carbon atoms.

> C  Which family of hydrocarbons is propene from?
>
> D  A butene molecule has four carbon atoms. Use the general formula of alkenes to predict the number of hydrogen atoms in the molecule.

### Key words

cracking, catalyst, unsaturated hydrocarbon, bromine water, alkene

> A  Identify the conditions needed for cracking reactions.
>
> B  Give an economic benefit to oil companies of cracking reactions.

## Exam tip

✔ Remember – cracking breaks down big molecules to produce smaller molecules of alkanes and alkenes.

## Questions

1  Name two possible products produced by cracking octane.

2  What is a catalyst?

3  Cracking produces hydrocarbons of two families. Name these families and give their general formulae.

4  A cracking reaction produces pentane ($C_5H_{12}$) and ethene ($C_2H_4$). Predict the formula of the starting material.

5  Suggest an environmental disadvantage of making products from chemicals obtained from crude oil.

↓ E
↓ C
↓ A*

▲ Many items we use every day are made of plastics

---

**A** List six items that contain plastics.

**B** Define the words polymer and monomer.

---

### Did you know...?

The polymers poly(ethene), PVC, and Teflon were all discovered by accident.

## Plastics everywhere

Look around you. How many things are made of plastics? Imagine life without plastics. Your great-great-grandparents probably did live without most of them – plastics only started to be widely used in the 1930s.

Most of the things we call plastics are made from polymers. But what's in a polymer?

**Polymers** are materials that have very big molecules. They are made by joining together thousands of small molecules, called **monomers**.

## Inside polythene

Polythene is an important polymer. Its properties make it useful. It is strong, flexible, and durable – perfect for bags and bottles.

Polythene molecules consist of thousands of atoms of carbon and hydrogen. The atoms are joined together in long chains.

The structure of polythene explains its properties:
* It is strong because the atoms in a molecule are joined together tightly, so it is difficult to break up a molecule.
* It is flexible because its molecules can slide over each other.

## Making polythene

Polythene molecules are made by joining together thousands of ethene molecules. Ethene molecules can join together because they have a double bond. The diagram shows how the molecules join together. Only a small part of the polythene molecule is shown.

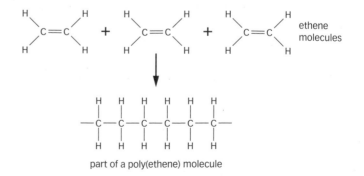

▲ Scientists call polythene **poly(ethene)**. The name shows that it is made from many – or poly – ethene molecules.

This is an example of a **polymerisation reaction**. You can use beads, paper clips, or molecular model kits to model polymerisation reactions.

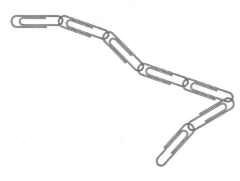

◀ Each paper clip represents a monomer molecule. The chain represents the polymer.

## More polymers

Poly(ethene) is not the only polymer. There are thousands of others. One of these is **poly(propene)**.

Poly(propene) is strong and rigid. It is not damaged by high temperatures. You can bend poly(propene) lots of times without it breaking. These properties mean that poly(propene) is useful for making many things, including:

- ropes
- underground water pipes
- dishwasher-safe food containers
- hinges for flip-top bottles.

The monomer used to make poly(propene) is propene. The formula of propene is $C_3H_6$. Its atoms are joined together like this:

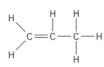

Thousands of propene molecules join together in long chains to make poly(propene). You can represent the reaction like this:

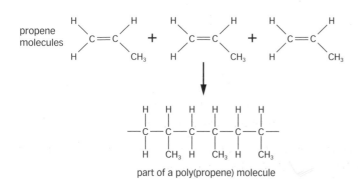

part of a poly(propene) molecule

**C** Match each of the uses of poly(propene) listed on this page to one or more properties that make it suitable for this purpose.

**D** Name the monomer used to make poly(propene).

**Exam tip**

✔ Practise writing equations to show how a polymer such as poly(propene) is made from its monomer.

### Questions

1 Name the monomer and polymer in the polymerisation reaction of ethene. ↓ E

2 Explain why alkene molecules can join together.

3 Write an equation to show how poly(propene) is formed from its monomer. ↓ C

4 Describe three benefits to people of the invention of polymers. These are social benefits.

5 Describe how you could use people to model a polymerisation reaction. ↓ A*

▲ HDPE makes rigid garden furniture

LDPE                    HDPE

▲ LDPE is less dense than HDPE because HDPE has fewer branches on its chains.

**A** Identify two properties that make HDPE more suitable than LDPE for making outdoor furniture.

**B** Explain why HDPE has a higher density than LDPE.

## Fit for purpose

Poly(ethene) and poly(propene) are just two of the many hundreds of synthetic polymers that chemists have created. Some polymers were discovered accidentally. Others were developed after hours of painstaking work in laboratories.

Each polymer has its own unique properties. The properties depend on

- what the polymer was made from
- the conditions under which it was made.

Different polymers have different uses. Their uses depend on their properties.

## Two types of poly(ethene)

You've probably used poly(ethene) bags. But did you know that you can get poly(ethene) garden furniture, too?

There are two types of poly(ethene):
- low density poly(ethene), **LDPE**
- high density poly(ethene), **HDPE**.

Each type of poly(ethene) has its own properties.

|  | LDPE | HDPE |
|---|---|---|
| Density (g/cm³) | 0.92 | 0.95 |
| Maximum temperature at which the polymer can be used (°C) | 85 | 120 |
| Strength (megapascal, MPa) | 12 | 31 |
| Relative flexibility | flexible | stiff |

LDPE and HDPE have different properties because their molecules have different structures.
- LDPE polymer molecules have side branches. The branches prevent the polymer molecules from lining up in a regular pattern, so the density is lower.
- HDPE polymer molecules have few side branches. Its molecules line up in a pattern, so the density is higher. The molecules are held together more strongly, so HDPE is stronger and has a higher melting point than LDPE.

# Dental polymers

Do you have fillings in your teeth? For years, dentists have filled teeth with a mixture of mercury, silver, and other metals. But this material conducts heat well, making it uncomfortable to eat very hot or very cold food.

Now, your dentist may offer you a white **dental polymer** filling. He will put a paste into your tooth and shine ultraviolet light on it. The ultraviolet light starts a reaction in which polymer molecules in the paste join together to make a solid filling.

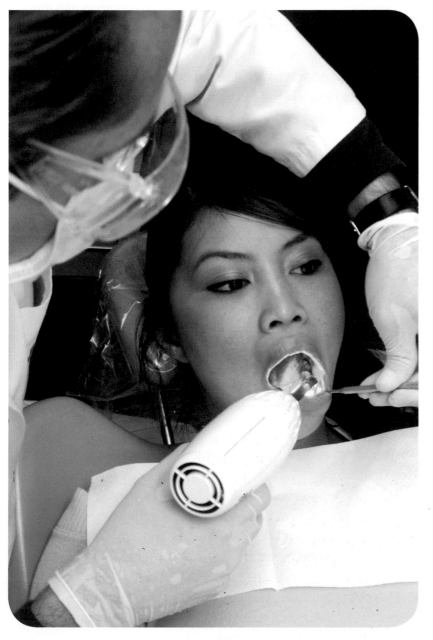

▲ Ultraviolet light starts a polymerisation reaction in the filling material

## Key words

LDPE, HDPE, dental polymer

## Exam tip

✔ Make sure you can explain how the properties of a polymer are linked to its uses. You do not need to remember details about specific polymers in your exams.

## Questions

1 Explain why different polymers have different properties. ↓ E

2 List one use each for LDPE and HDPE. Explain how the properties of each of these polymers make them suitable for these uses. ↓ C

3 Suggest two advantages of filling your teeth with a white polymer and not a mixture of metals.

4 Use ideas about particles to explain why HDPE is stronger and has a higher melting point than LDPE. ↓ A*

## Learning objectives

After studying this topic, you should be able to:

✔ explain how the properties of polymers determine their uses

▲ Disposable nappies contain hydrogels

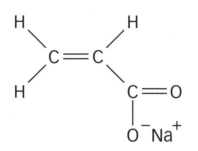

▲ The monomer from which poly(sodium propenoate) is made.

**B** Explain why the properties of shrink wrap make it a suitable material for packaging items such as DVDs.

## Key words

hydrogel, shape memory polymer, smart material, breathable material

## Perfect polymers

Chemists continue to create new polymers with perfect properties for particular purposes. These include waterproof coatings for fabrics, wound dressings, and shape memory polymers.

## Hydrogels

Disposable nappies absorb huge amounts of urine – and the baby doesn't even feel wet. How do they do this?

Disposable nappies contain **hydrogels**. Hydrogels are made from polymers such as poly(sodium propenoate).

Normally, the polymer chains are coiled up. But if you take away all the sodium particles, the chains uncoil. Water molecules are attracted to the uncoiled chains, and the hydrogel absorbs up to 500 times its own weight of water.

Hydrogels also make excellent wound dressings. The hydrogel protects the wound from infection and controls bleeding. Because hydrogels do not stick to the skin, you can remove them easily without damaging the skin.

**A** Explain why the properties of hydrogels make them suitable for use in disposable nappies.

## Shape memory polymers

If you've bought a DVD recently, you might have noticed that it was shrink-wrapped. Shrink wrap is a **shape memory polymer**. Shape memory polymers are **smart materials** – they change in response to their environment.

Shrink wrap is made from polymers such as poly(ethene). Here's how:

- Heat the poly(ethene) until it becomes a thick liquid. The molecules are now coiled together randomly.
- Cool the liquid quickly. As it cools, quickly stretch out the polymer to make a thin film of solid. The molecules are now stretched out.
- Heat the thin film. The stretched molecules suddenly return to their coiled shape. The film shrinks and wraps tightly round the DVD case.

# Waterproof clothing

Ben loves walking, but not in the rain. His waterproofs are made of nylon. Nylon stops water getting in. It is also tough and lightweight. But nylon stops water vapour getting out. So Ben's sweat condenses inside his waterproofs, making him feel damp and uncomfortable. Yuk!

Suzette's waterproofs are much more comfortable. They are made of a **breathable material**. The material contains three layers. The middle layer is made of a polymer called PTFE. This has many tiny holes. Each hole is too small for water droplets to pass through, but big enough for water vapour to pass through. So rain can't get in, but water vapour from sweat can get out.

▲ Breathable waterproof coats

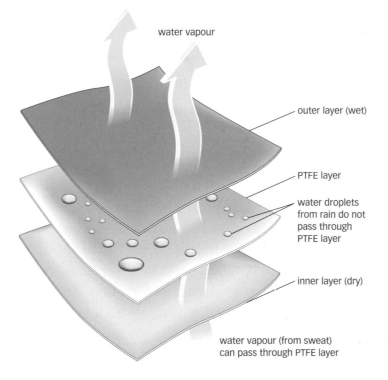

water vapour

outer layer (wet)

PTFE layer

water droplets from rain do not pass through PTFE layer

inner layer (dry)

water vapour (from sweat) can pass through PTFE layer

▲ Breathable fabrics are made up of materials with different properties

## Did you know...?

The PTFE layer in breathable materials has around 1.4 billion holes per square centimetre.

C Explain why breathable materials stop water getting in but allow sweat to get out.

## Exam tip

✓ You do not need to remember the names of polymers used as hydrogels, in shape memory polymers, or in waterproof clothing. But be prepared to answer questions about polymers and their properties from information given in the exam.

## Questions

1 List two uses of hydrogels.

2 Give one use of shape memory polymers.

3 Make a table showing the uses of three types of polymer and explaining how their properties make them suitable for these purposes.

↓ E

↓ C

## Key words

non-biodegradable, landfill site

▲ Waste plastic litters this beach

## Did you know...?

It has been estimated that there are around 17 000 pieces of plastic per square kilometre of ocean, and that plastic bags take 1000 years to degrade.

▲ Landfill sites waste valuable land

## Litter, litter, everywhere

Many polymers are unreactive. They do not react easily with acids, alkalis, or many other chemicals. They are often tough and strong. These properties are some of the reasons why polymers are so very useful. For example, we can store shampoo and water in plastic bottles without worrying that the bottle will dissolve. We can bury plastic gas pipes deep underground, knowing that they will serve us for many years. Unfortunately, these very benefits make polymers difficult to dispose of.

Decay bacteria help substances rot away. But many polymers are **non-biodegradable**. They cannot be decomposed by bacterial action. Plastic waste litters the Earth almost everywhere.

A What does non-biodegradable mean?

B Explain why many polymers are difficult to dispose of.

## Disposing of polymers

When you throw out a used plastic item, the chances are it will end up in a **landfill site**. These are places where local authorities take waste to be buried. In the UK, around 85% of waste is dumped in landfill sites. They eventually become full of waste. It is difficult to find land for a new landfill site. Local people may object to it, for example because it will be unsightly. Although the plastic itself may not smell, other materials rotting away are smelly and may attract rats and gulls.

Crude oil is the raw material for most polymers. It is also the source of fuels such as petrol. Waste plastic can be burnt to release heat energy. This can be used to generate electricity or to heat buildings. The plastic must be burnt at a high temperature to stop toxic gases being made. Unless the heat energy released is used, disposal by burning is a waste of a valuable resource.

Waste plastic can be burnt in incinerators like this

It is possible to recycle polymers. For example, PET is the tough plastic used to make drinks bottles. It can be recycled to make fibres for clothing. Unfortunately, plastic waste is usually a mixture of different polymers. It must be sorted into separate types of polymer, often by hand. This is difficult and expensive to do.

## Chemists to the rescue

Chemists are developing ways to make addition polymers such as poly(ethene) biodegradable. Starch can be added to the polymer during manufacture. Bacteria can break down starch once the polymer gets wet. This causes the plastic item to break down into very small pieces. It has not rotted away, but it is no longer litter.

Bags are also now being made of cornstarch, a material obtained from maize. Cornstarch is biodegradable. However, some people think it is not ethical to grow maize to make bags when some people in the world do not have enough to eat.

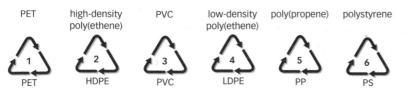

Recycling symbols found on plastic items

**Exam tip** **AQA**

✔ Make sure you can describe the drawbacks of different ways to dispose of waste polymers.

**Did you know…?**

It takes 25 PET bottles to make enough fibre for a fleece jacket.

**Questions**

1 Describe three ways in which waste polymers can be disposed of.

2 Explain why recycling symbols are put on plastic items.

3 Describe the benefits of biodegradable polymers.

4 Explain why burying waste plastics or burning them is a waste of valuable resources.

5 In 2002, Ireland introduced a tax on plastic carrier bags. Sales of carrier bags fell by 90%, but sales of other bags increased by 400%.

(a) Suggest why the tax was introduced.

(b) Suggest why plastic bag sales changed in the way they did.

↓ E

↓ C

↓ A*

▲ This person has consumed too much ethanol

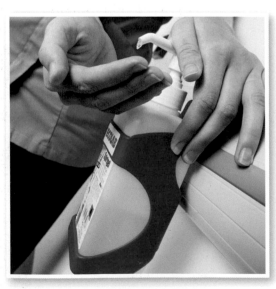

▲ Ethanol is an effective disinfectant

## Using ethanol

Ethanol is big business. It makes huge profits for the companies that produce it. But its costs are huge, too. In 2005 ethanol was responsible for around 26 600 deaths in England and Wales. Ethanol costs the National Health Service an estimated £3 billion each year.

So what is this stuff called ethanol? Its formula is $C_2H_5OH$. It is the main ingredient in alcoholic drinks. It relaxes people and makes them lose their inhibitions.

Ethanol is not just used in drinks. It has many other uses, including:

- as a solvent in perfumes and after-shave
- as a disinfectant in hospital hand gels
- as a fuel for cars (see spread C1.16).

> **A** List four uses of ethanol.
>
> **B** Describe one problem associated with ethanol.

## Making ethanol

There are two ways of making ethanol. People have been using one of these methods – **fermentation** – for many centuries. The other method was developed more recently.

### Ethanol from ethene

The cracking of crude oil fractions produces huge amounts of ethene gas. Chemical companies react some of this ethene with steam to make ethanol. A catalyst, usually phosphoric acid, speeds up the reaction. The reaction works well at about 300 °C. These equations represent this **hydration** reaction:

$$\text{ethene} + \text{steam} \rightarrow \text{ethanol}$$
$$C_2H_4(g) + H_2O(g) \rightarrow C_2H_5OH(g)$$

The process is efficient because there are no waste products. It is a continuous process – it carries on for as long as ethene and steam are being added. However, ethene is a non-renewable resource. It is made from crude oil, which will run out one day.

Ethanol made from ethene is not normally used in alcoholic drinks. It can be used as a fuel and as a solvent.

## Ethanol from sugars

Wine is made from grapes by fermentation. The grapes contain sugars, including **glucose**. Enzymes in yeast break down the glucose into ethanol and carbon dioxide. The ethanol mixes with other chemicals in the grapes to make wine. Fermentation works best at a temperature of 37 °C. The yeast enzymes are a **natural catalyst**. The raw materials for fermentation can be grown again, so they are renewable.

$$\text{glucose} \rightarrow \text{ethanol} + \text{carbon dioxide}$$
$$C_6H_{12}O_6(aq) \rightarrow 2C_2H_5OH(l) + 2CO_2(g)$$

Ethanol for fuel is often made by the fermentation of crops such as sugar cane.

> C  List the temperatures and catalysts for the two methods of making ethanol.

## Making ethanol – which way is better?

Which is the better way of making ethanol? Each method has its pros and cons.

The crude oil that makes ethene is non-renewable

It is not right to use land to grow plants for ethanol fuel. The land is better used for crops.

I work in a fermentation factory. Making ethanol by fermentation provides many more jobs than making ethanol from ethene.

Of the two processes, fermentation needs less energy because it happens at a lower temperature

▲ The ethanol in cider is made from sugars in apples

### Did you know...?

Wasps can get drunk in autumn when they feed on fermented sugars from rotting apples.

### Exam tip

✔ Remember the advantages and disadvantages of making ethanol by the two methods.

### Questions

1  Name the raw materials for making ethanol from ethene. ↓E

2  Name the waste product formed when ethanol is made by fermentation.

3  Draw up a table to compare the advantages and disadvantages of making ethanol by the two methods. ↓C

4  Write a balanced symbol equation to summarise the two methods by which ethanol can be produced. ↓A*

### Learning objectives

After studying this topic, you should be able to:

✔ evaluate the impacts of using vegetable oils in foods

## Edible oils

What's the connection between the field of flowers below, and the bus and the plate of chips on the left?

The flowers produce rapeseed oil that fuels the bus and cooked the chips. Rapeseed oil is just one of many types of oil obtained from plants. In 2008, humans used more than 18 million tonnes of this oil.

## Oil matters

Plant oils are important. We heat oils such as sunflower oil, rapeseed oil, and soya bean oil and use them to cook food. Oils are used for cooking because they have higher boiling points than water. So foods cook more quickly in oils. Oil-cooked foods have different flavours from those cooked in water – think about the difference in flavour between boiled potatoes and chips, for example.

Sunflower, rapeseed, and soya bean oils are particularly suitable for frying because they have high flash points. This means they catch fire only at high temperatures. You need to heat sunflower oil to 274 °C before it will set itself alight, for example.

People add olive oil to salads because they enjoy its flavour. Plant oils are vital ingredients in many types of pastries and biscuits, too. They help give these foods their crumbly texture.

**A** List three ways of using plant oils in food and cooking.

**B** Explain why sunflower oil is suitable for frying.

# Oil benefits

Oils are high in energy. If you eat lots of oil-rich food, you may put on weight. On the other hand, if you're planning an expedition to the North Pole, you may decide to take with you foods that are high in oil. Your body needs extra energy from food in cold climates or if you are physically active.

The table shows the energy content of sunflower oil compared to other foods.

| Food | Energy in 100 g of the food (kJ, approximate values) |
|------|------------------------------------------------------|
| sunflower oil | 3800 |
| potato crisps | 2100 |
| apple | 200 |
| chicken breast | 600 |

> **C** List the foods in the table according to how much energy they provide.
>
> **D** Suggest why potato crisps have a high energy content.

Plant oils provide **nutrients** as well as energy. Sunflower oil is high in vitamin E. There is evidence that vitamin E is an antioxidant. This means that it protects cells membranes from damage and may help prevent cancer.

Olive oil contains a compound called oleocanthal. Scientists have found that oleocanthal has anti-inflammatory properties. They suggest that the compound may explain why there is less heart disease in countries where people eat lots of olive oil.

Some plant oils are a good source of omega-3 fatty acids. Many scientists have researched the benefits of these substances. They found evidence that omega-3 fatty acids may help to prevent cancer, heart disease, and even poor behaviour.

## Did you know...?

Scientists recently discovered that early artists used paints based on walnut oil and poppy seed oil to create the earliest known painting in a cave in Afghanistan. The painting dates from about 650 AD.

## Key words

**nutrient**

## Exam tip

✔ Remember – plant oils can come from fruits, seeds, and nuts. They provide energy and nutrients.

## Questions

1 List three plants that provide oils.

2 Describe three health benefits of plant oils.

3 Explain the benefits of cooking foods in plant oils.

4 Describe one way in which plant oils may harm health.

5 Write a paragraph to evaluate the impacts of plant oils on health.

# 25: Oils from fruit and seeds

## Learning objectives

After studying this topic, you should be able to:

✔ describe how plant oils are extracted from fruits, seeds, and nuts

▲ Sunflower oil is extracted from sunflower seeds

▲ A traditional olive oil press

## Where do plant oils come from?

Plants store oils in their fruits, seeds, and nuts. The oil is part of a seed's food store. The seed needs the oil as an energy source when it starts to germinate.

▲ Olive oil is extracted from the fruit of olive trees

▲ Peanut oil is extracted from peanuts

How do we extract oils from fruits, seeds, and nuts? Read on to find out.

## Extracting plant oils

### Pressing

People have been extracting oil from olives for centuries. Roman olive presses still exist in Morocco. Today, traditional olive oil producers continue to use crushing and **pressing** to separate oil from the substances it is mixed with in olives. Here's how:

- Crush the olives into a paste to release oil from the cell vacuoles.
- Mix the paste for 30 minutes so that small oil droplets join together to form bigger ones.
- Press the paste by spreading it onto fibre discs and applying pressure to squash out the liquids.
- Collect the liquid mixture – of mainly oil and water – and allow it to settle. The oil floats on the water and is poured off.
- Remove impurities from the olive oil.

## Solvent extraction and distillation

Extracting oil from sunflower seeds is less straightforward. The process has several stages:

- Remove the hulls from the seeds.
- Press the seeds to obtain some oil.
- Add a solvent, such as hexane, to the solid that remains. Any oil still mixed with the solid will dissolve.
- Use **distillation** to separate the solvent from the oil that is dissolved in it.

> **A** Explain how olive oil is separated from olives by crushing and pressing.
>
> **B** Explain how sunflower oil is extracted from sunflower seeds in a multi-step process.

## Steam distillation

You can extract lavender oil by **steam distillation**. Here's how.

Set up the apparatus like this.

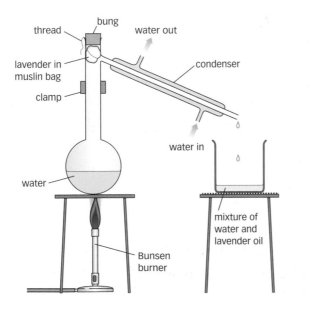

- Heat the water so that it boils and begins to make steam.
- Allow the steam to pass through the plant material. The volatile oils in the lavender vaporise.
- The steam and plant oil vapour move to the condenser. Here, they cool down and become liquid.
- The beaker contains lavender oil and water.

**Exam tip**

- ✔ Remember – olive oil is extracted by pressing and sunflower oil is extracted by pressing, solvent extraction, and distillation. You do not need to learn the details of these processes off by heart.

**Questions**

1 Name one fruit, one nut, and one seed from which oils are extracted. ↓ E

2 List the stages used to extract sunflower oil from sunflower seeds. ↓ C

3 Identify the changes of state that happen in the steam distillation process. Write down the stages of the process at which each change of state occurs. ↓ A*

# 26: Emulsions

## Learning objectives

After studying this topic, you should be able to:

✓ explain how the properties of emulsions make them suitable for their uses

✓ explain how emulsifiers work

▲ What do these three substances have in common?

**A** Describe what happens when you mix oil and water.

**B** Explain why oil and water don't mix well.

## Salad dressing, paint, and ice cream

What do salad dressing, ice cream, and emulsion paint have in common?

They are all emulsions. But what are emulsions, and what are they like inside? Read on to find out more.

## Mixing oil and water

Vegetable oils and water don't mix together well. However hard you shake them together, the liquids separate out when you stop shaking. Oils are less dense than water, so they float on its surface.

Oil and water don't mix because their particles are too different. Vegetable oil molecules include long hydrocarbon chains. These chains cannot interact with small water molecules.

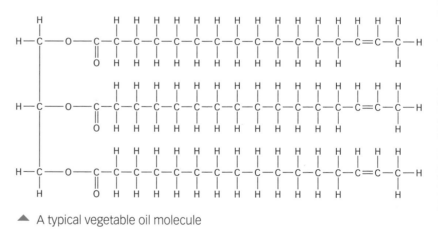

▲ A typical vegetable oil molecule

## Inside emulsions

If you add an **emulsifier** to oil and water and shake well, the oil and water no longer separate out. The emulsifier stabilises the mixture and an **emulsion** forms.

Emulsions are more viscous, or thicker, than the liquids they are made from. This property leads to emulsions having a wide range of uses:

• Salad dressing contains vinegar, oil, and an emulsifier. Its high viscosity means it coats lettuce leaves well.

• Ice cream is an emulsion. The properties of the emulsion contribute to the special texture and appearance of ice cream.

• Emulsion paint coats walls well because of its particular texture and high viscosity.

- Many cosmetics are emulsions. Hand cream, moisturising cream, and shaving cream are oil-in-water emulsions. They consist of tiny oil droplets dispersed in water.
- More greasy cosmetics, such as sunscreens, are usually water-in-oil emulsions. They are made up of water droplets dispersed in oil.

## Emulsifier safety

You've probably heard of some emulsifiers. Mayonnaise, for example, contains egg. Other emulsifiers are less well-known.
- Carrageenan (E407) is an emulsifier extracted from red seaweed. It is an ingredient of many processed desserts and ice cream.
- Tragacanth (E413) is a gum obtained from the sap of a plant that grows in Iran. It is used in cake decorations.

Artificial emulsifiers are identified by **E-numbers**. Additives with E-numbers have been safety tested and licensed by the European Union.

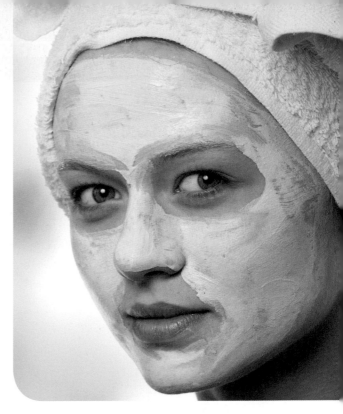

▲ Many cosmetics are emulsions

### Emulsifiers at work

Emulsifier molecules have two different ends:
- One end interacts well with water molecules. This is the **hydrophilic** end.
- The other end interacts well with oil molecules, and badly with water molecules. This is the **hydrophobic** end.

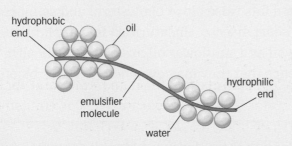

In an oil-in-water emulsion, emulsifier molecules coat the surface of oil droplets. The hydrophobic ends of the emulsifier molecules interact with the oil droplets. The hydrophilic ends interact with the water. These coatings keep the oil droplets evenly dispersed throughout the emulsion, and stop them clumping together to form their own separate layer.

### Key words

emulsifier, emulsion, E-number, **hydrophilic**, **hydrophobic**

### Exam tip

✔ Remember that emulsifiers stop emulsions separating out.

### Questions

1  Describe the property of emulsions that makes them suitable for salad dressings.

2  Explain why plants make and store oils.

3  Explain how emulsifiers stop oil and water separating out in emulsions.

▲ Bread with olive oil …

▲ … and toast with butter

## Spread for your bread

Riana likes butter on her toast. Matthew prefers margarine. Donnatella drizzles olive oil over bread for a delicious snack. Butter, margarine, and olive oil provide similar amounts of energy. They are all mixtures of compounds consisting of atoms of carbon, hydrogen, and oxygen. What makes them different?

## Liquid or solid?

Butter and margarine are solid at room temperature. Most plant oils are liquid at room temperature. They melt at lower temperatures than butter and margarine.

The structures of the molecules in fats and oils explain their different melting temperatures.

- Butter is a **saturated** fat. There are no double bonds in its molecules.
- Most plant oils have **unsaturated** hydrocarbon chains in their molecules. The hydrocarbon chains contain double carbon–carbon bonds.
  1. **Monounsaturated** fats have one double bond per hydrocarbon chain.
  2. **Polyunsaturated** fats have several double bonds per hydrocarbon chain.

## Good for your health?

Saturated fats raise blood cholesterol and so increase the risk of heart disease. Unsaturated fats in plant oils such as olive oil and sunflower oil are better for health. There is evidence that they may even help to lower blood cholesterol.

> A Explain the difference between saturated fats and unsaturated fats.
>
> B Explain why sunflower oil is better for your health than butter.

# Detecting double bonds

Achita has three plant oils. She needs to know if any of them are high in saturated fats.

Achita adds orange bromine water to samples of the oil. The test is the same as that for detecting double bonds in alkenes (see spread C1.18). Here are Achita's results.

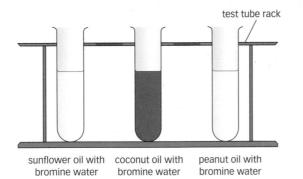

sunflower oil with    coconut oil with    peanut oil with
bromine water         bromine water       bromine water

The bromine has reacted with sunflower oil to form a colourless mixture. This shows that sunflower oil is unsaturated. Bromine atoms have added to the carbon atoms on either side of the double bonds.

There is no change in the coconut oil test tube. Bromine has not reacted with the oil. So coconut oil is saturated. It contains no carbon–carbon double bonds.

## Making margarine

Food companies harden unsaturated vegetable oils by adding hydrogen gas to them. These reactions happen at about 60 °C. A nickel catalyst speeds up the reaction.

In these **hydrogenation** reactions, hydrogen atoms add to carbon atoms on both sides of the double bonds.

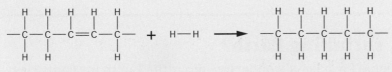

▲ The diagram shows just part of a plant oil molecule

Hydrogenation reactions convert unsaturated oils into saturated ones. The saturated oils have higher melting points. They are solid at room temperature. They can now be used as spreads and to make cakes and pastries.

## Key words

saturated, unsaturated, monounsaturated, polyunsaturated, hydrogenation

**C** Explain what is shown by the result of the test of bromine water with peanut oil.

## Exam tip

✓ Molecules in saturated oils have no double bonds. Unsaturated oil molecules do have double bonds. Bromine water tests for unsaturation.

## Questions

1 What is a saturated fat?

2 Describe how to test an oil for unsaturation. Explain what different results would mean.

3 Describe the conditions needed for hydrogenation reactions.

4 Predict whether butter or olive oil is higher in unsaturated fats. Explain your decision.

E

C

A*

# 28: Inside the Earth

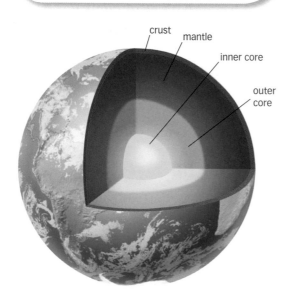

▲ The structure of the Earth

▲ A model for the Earth?

## The structure of the Earth

Think about all the things you've used today. Where did they all come from? The answer is the Earth, its **atmosphere**, and the oceans. All the minerals and other resources humans use come from one of these three sources.

The Earth is a sphere. It is made up of several layers. The main ones are:

- a rocky **crust**, which is thin compared to the other layers
- the **mantle**, which goes down almost halfway to the centre of the Earth; the mantle has solid properties, but can flow very slowly
- the **core**, which is made of iron and nickel.

Surrounding the Earth is a mixture of gases. This is the atmosphere.

The radius of the Earth is about 6370 km. Imagine you could ride in a jumbo jet to the centre of the Earth. The journey would take about seven hours at top speed, and there would be just enough fuel to return to the surface. But it is not really possible to study the Earth's structure directly. The deepest mine is only 3.5 km deep.

> A Describe the structure of the Earth, starting at its centre.
>
> B Imagine the Earth as an egg. Which parts of the Earth can be represented by the egg shell, the white, and the yolk?
>
> C Explain why it is not possible to study the Earth's structure directly.

## A shrinking Earth?

People have been wondering about the Earth for centuries. Where did mountains come from? Why are continents the shapes they are?

Scientists once believed that the features of the Earth's surface were the result of the shrinking of the crust as the Earth cooled down after it was formed. Valleys and mountains were like the wrinkles on the surface of a drying apple. However, scientists have since collected evidence to show that this theory must be false.

## Controversial theory

In 1912, the German scientist Alfred Wegener put forward a new theory to explain the history of the Earth. He suggested that the continents were once joined together, and that they had gradually moved apart. Wegener supported his theory with evidence:

- The shapes of Africa and South America look as if they might once have fitted together.
- Fossils of the same plants had been found in both Africa and South America.
- The two continents have the same rock types at the edges where they might have been joined.

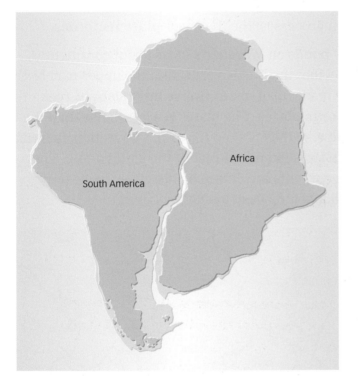

South America

Africa

At the time, most other scientists did not accept Wegener's theory of **crustal movement**, or **continental drift**. They could not see how the continents might have moved. They were also reluctant to believe Wegener because he was not a geologist.

However, by the 1950s, scientists were beginning to support Wegener's theory. They discovered evidence suggesting that the Atlantic Ocean sea floor was spreading. They discovered convection currents in the Earth's mantle. And they discovered similar rocks and fossils between different pairs of continents.

### Did you know...?

The deepest hole ever drilled is the Kola Superdeep Borehole in Russia. It reached 12.261 km through the crust in 1989. By then it was too hot for drilling to continue.

### Key words

atmosphere, crust, mantle, core, crustal movement, continental drift

### Questions

1 List two parts of the Earth's structure which are solid.

2 Describe the Earth's mantle. ↓ E

3 Describe Wegener's theory of continental drift.

4 List three pieces of evidence that support Wegener's theory of continental drift. ↓ C

5 Explain why scientists were reluctant to accept Wegener's theory at first. ↓ A*

▲ Haiti disaster

## Restless Earth

12 January 2010: A catastrophic magnitude 7.0 **earthquake** strikes Haiti. By 24 January, at least 52 aftershocks had been recorded. A quarter of a million people die and the same number are injured. A million more lose their homes.

23 April, 1902: Mount Pelée **volcano** in Martinique begins to erupt. More than 29 000 people die in its ash flows.

What caused these disasters? Could they have been predicted?

## Tectonic plates

One big idea explains most volcanoes, earthquakes, and other Earth processes: the theory of **plate tectonics**.

The theory builds on Wegener's ideas about continental drift. It says that the Earth's crust and the upper part of the mantle are made up of about twelve huge slabs of rock called **tectonic plates**. The plates are less dense than the mantle below, so they rest on top of it. Convection currents deep within the solid mantle carry the plates along. The convection currents are driven by heat from radioactive processes happening deep inside the Earth.

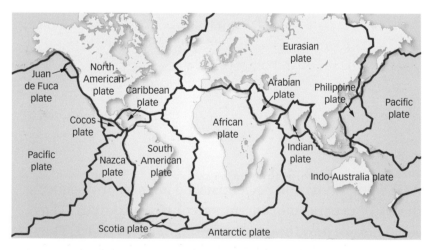
▲ Tectonic plates

Tectonic plates move very slowly, at a speed of a few centimetres a year. Global positioning satellites track their movements.

A  What is a tectonic plate?

B  Describe how tectonic plates are carried along.

## Earthquakes

Earthquakes happen when tectonic plates move against each other suddenly.

At the San Andreas Fault in California, USA, two enormous tectonic plates are moving past each other in opposite directions. But, because of friction, the plates cannot slide smoothly and they sometimes get stuck. Huge forces build up as they keep trying to pass each other. Eventually the two plates overcome the frictional forces. They slip suddenly. There is an earthquake.

The place where the plates slip is the **focus** of an earthquake. Shock waves spread from the focus in all directions. The waves make buildings collapse. In earthquakes, falling buildings may kill and injure thousands of people.

Earthquakes are common at all moving plate boundaries. Undersea earthquakes may cause enormous waves – or tsunamis – which do great damage when they reach land.

Scientists cannot predict exactly when plates will suddenly slip. So people living on plate boundaries can expect an earthquake at any time.

## Volcanoes

A volcano is a vent in the Earth's crust from which magma (liquid rock), ash, and gases such as carbon dioxide and sulfur dioxide erupt. Every year there are around 50 volcanic eruptions.

Scientists monitor active volcanoes closely to look for signs that an eruption may happen soon. A volcano may be about to erupt if

- there are earthquakes nearby
- the shape of the volcano changes
- the volcano gives off more gases than usual.

But scientists can still never be sure exactly when a volcano will erupt, or how severe eruptions will be.

## Making mountains

Mountains may form when tectonic plates collide. For example, when two plates push against each other, rocks at the edge of one of the plates buckle and fold to form a mountain chain.

▲ San Andreas Fault

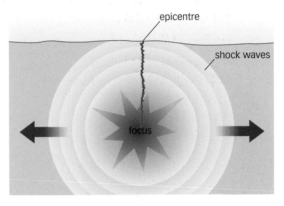

▲ Focus, epicentre, and shock waves of an earthquake

C  Explain why scientists cannot predict exactly when an earthquake will happen.

D  Describe the signs scientists look for to predict volcanic eruptions.

### Questions

1  List two types of Earth events that can be explained by plate tectonics.  ↓ E

2  Describe how plates sliding past each other may cause earthquakes.  ↓ C

3  Suggest two actions a local council might take if scientists predict a volcanic eruption.  ↓ A*

## Learning objectives

After studying this topic, you should be able to:

✔ describe the composition of the Earth's atmosphere

✔ know how to separate gases from the air

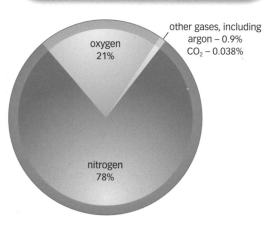

other gases, including
argon – 0.9%
$CO_2$ – 0.038%

oxygen
21%

nitrogen
78%

▲ The pie chart shows the proportions of the main gases in the air

**A** Make a table to show the proportions of the main gases in the air.

## Breathing on the Moon

July 1969. Humans land on the Moon. Astronauts Neil Armstrong and Buzz Aldrin wear breathing apparatus for the first ever Moon walks. Why?

Unlike the Earth, no gases surround the Moon. The Moon has no atmosphere.

## The atmosphere today

The Earth's atmosphere is a mixture of gases surrounding the planet. The Earth's gravity stops these gases escaping into space. The lower part of the atmosphere is the air that we breathe. It has stayed much the same as it is today for the last 200 million years.

Just two gases, nitrogen and oxygen, make up about 99% of the air. They are both elements. There are smaller proportions of other gases in the air, including carbon dioxide, water, and noble gases such as argon.

## Gases from the atmosphere

The atmosphere is a vital source of raw materials for many processes. Nitrogen makes ammonia for fertilisers. Hospitals use oxygen to treat patients. The noble gases have a great variety of uses. But how do companies separate the gases of the air?

Nitrogen, oxygen, and the other gases of the air have different boiling points.

| Gas | Boiling point (°C) |
|---|---|
| nitrogen | −196 |
| oxygen | −183 |
| argon | −186 |

**B** Make a bar chart to show the boiling points of the gases in the table.

Their different boiling points mean that the gases can be separated. Here's how.

- Cool the air to – 200 °C in stages, so that its gases condense to form a mixture of liquids. As the air cools:
  - water vapour condenses and is removed
  - carbon dioxide freezes at –79 °C and is removed
  - oxygen and nitrogen condense at their boiling points.
- Separate the mixture of liquid oxygen, liquid nitrogen, and liquid argon by fractional distillation.

▲ Liquid nitrogen is used to freeze food

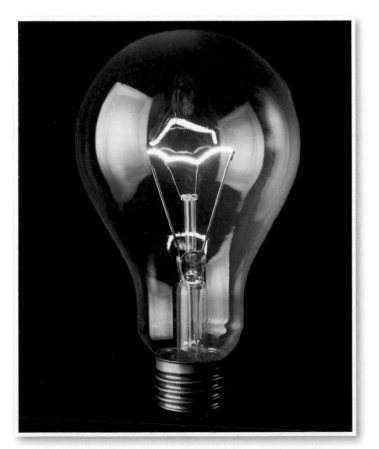

▲ Filament light bulbs are filled with argon gas. They are being phased out because they are not energy efficient.

### Did you know...?
Foods such as potato crisps are packaged in nitrogen gas to increase their shelf life.

▲ Hospitals use oxygen to treat patients

## Questions

1 List the gases in the atmosphere and give their proportions.

⬇ E

2 Name three elements and two compounds that are present in the atmosphere.

⬇ C

3 Describe how nitrogen and oxygen are obtained from the atmosphere.

4 Give one use each for nitrogen, oxygen, and argon.

⬇ A*

## Different theories

The Earth's atmosphere hasn't always been as it is today. So where has our atmosphere come from? How was it formed? Scientists have several theories to explain the origins of the atmosphere. We cannot know which, if any, is correct. No one was around to record events as they happened. One possible theory involves plants and volcanoes.

## The first billion years

The Earth is about 4.5 billion years old. There was a lot of **volcanic activity** during its first billion years. When volcanoes erupt, they release huge quantities of gases. These are mainly water vapour and carbon dioxide, with smaller amounts of other gases such as ammonia ($NH_3$) and methane ($CH_4$).

It is likely that the Earth's early atmosphere came from gases released by volcanoes from inside the Earth's crust. The water vapour condensed to form the oceans. So the early atmosphere was mainly carbon dioxide with small proportions of ammonia and methane. There would have been little or no oxygen.

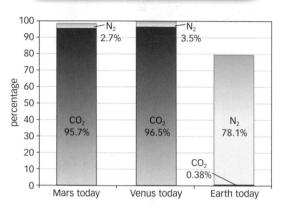

▲ The Earth's early atmosphere was probably like the present day atmospheres of Mars and Venus

> **A** Describe the ways in which the Earth's early atmosphere was like the present day atmospheres of Mars and Venus.

> **B** Outline the main differences between the Earth's early atmosphere and its atmosphere today.

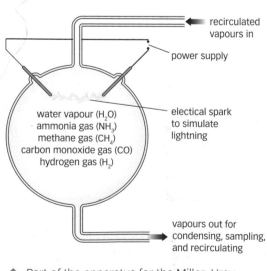

▲ Part of the apparatus for the Miller–Urey experiment

## Life begins

Billions of years ago, life began. No one knows exactly how, but many scientists have collected evidence and suggested theories.

One theory, the **primordial soup theory**, suggests that gases in the early atmosphere reacted with each other in the presence of sunlight or lightning to make complex molecules that are the basis of life.

In 1953, two scientists designed an experiment to test the theory. The **Miller–Urey experiment** simulated a lightning spark in a mixture of the gases of the early atmosphere. A week later, more than 2% of the carbon in the system had formed compounds from which proteins in living cells are made. This, said the scientists, supports the primordial soup theory.

# A changing atmosphere

Plants make their own food by **photosynthesis**. They take carbon dioxide from the atmosphere and release oxygen gas:

carbon dioxide + water → oxygen + glucose

As plants evolved from early living organisms, their photosynthesis reduced the amount of carbon dioxide in the atmosphere. Photosynthesis also increased the proportion of oxygen in the atmosphere until it reached today's level.

Photosynthesis was not the only reason why carbon dioxide levels decreased.

## Locking up carbon in rocks

Carbon dioxide is a soluble gas, and large amounts of it dissolved in the oceans.

Shellfish and other sea creatures used some of this carbon dioxide to make their shells and skeletons. When the animals died, they fell to the bottom of the ocean. After many years, limestone, a **sedimentary rock**, formed from their shells and skeletons. The carbon atoms were locked away in limestone, mainly as calcium carbonate.

## Locking up carbon in fossil fuels

Millions of years ago, dead plants and animals decayed under swamps. The dead organisms formed fossil fuels. The carbon atoms of the plants and animals were locked up in underground stores of coal, oil, and gas.

▲ Large amounts of carbon dioxide dissolve in the oceans

▲ Limestone rock was formed from the shells and skeletons of shellfish and other sea creatures

## Questions

1. Name four gases that were probably present in the Earth's early atmosphere. Where did these gases come from? ↓ E

2. Explain three ways by which carbon dioxide was removed from the Earth's early atmosphere.

3. Explain the origin of atmospheric oxygen. ↓ C

4. Explain why we cannot be sure how the Earth's atmosphere was formed.

5. Describe the primordial soup theory of the origin of life, and the experimental evidence that supports it. ↓ A*

### Did you know...?

At current rates of photosynthesis it would take living things just 2000 years to make all the oxygen found in the Earth's atmosphere today.

### Exam tip  AQA

✓ Remember that oxygen increased over time. Carbon dioxide decreased.

## Learning objectives

After studying this topic, you should be able to:

✔ explain and evaluate the effects of human activities on the atmosphere

## Key words

**respiration, combustion, reservoir, deforestation**

▲ Plants remove carbon dioxide from the atmosphere

## Exam tip

✔ Photosynthesis and dissolving in the oceans remove carbon dioxide from the atmosphere. Respiration and combustion add to atmospheric carbon dioxide.

## Carbon dioxide – on the up

Carbon dioxide is vital to life. Without it, plants could not make their own food. Animals – including humans – would have nothing to eat. And Earth would be too cold for life as we know it.

Since about 1800, the percentage of carbon dioxide in the atmosphere has been increasing year by year. Why? And why does it matter?

## Into and out of the atmosphere

### Into the atmosphere

Two main processes add carbon dioxide to the atmosphere:

- **Respiration** is the process by which plants and animals release energy from food. Carbon dioxide is one of the waste products of this process.

    glucose  +  oxygen  →  carbon dioxide  +  water

- **Combustion** also releases carbon dioxide into the atmosphere. Fossil fuels contain carbon atoms that have been locked away for millions of years. When they are burnt, they join with oxygen atoms and enter the atmosphere as carbon dioxide molecules. For example:

    methane  +  oxygen  →  carbon dioxide  +  water

### Leaving the atmosphere

Today, about 0.038% of the Earth's atmosphere is carbon dioxide. Plants remove some of this gas to make their own food by photosynthesis. Carbon dioxide is removed from the atmosphere when it dissolves in the oceans, too.

## The carbon cycle

The carbon cycle summarises the processes that add and remove carbon dioxide from the atmosphere.

The carbon cycle also shows the stores – or **reservoirs** – of carbon, including the atmosphere, the oceans, sedimentary rocks, and fossil fuels.

> **A** Name two processes that add carbon dioxide to the atmosphere.
>
> **B** Name two processes that remove carbon dioxide from the atmosphere.

## Off balance

You have just seen that some processes add carbon dioxide to the atmosphere and some remove it. If these processes balance, then the concentration of carbon dioxide in the atmosphere will not change.

Since 1800, these processes have not been balanced. Human activities have been adding carbon dioxide to the atmosphere faster than it is removed:

- People burn fossil fuels to generate electricity, heat houses, and fuel vehicles.
- The human population is increasing. This adds to the demand for energy and leads to more fossil fuels being burnt. It also adds to the demand for land for buildings, roads, and farms.
- Forests are cut down, or burnt, to make way for human use. This **deforestation** has a big impact on carbon dioxide levels. When trees burn, they release carbon dioxide to the atmosphere. Fewer trees remain to remove carbon dioxide from the atmosphere by photosynthesis.

As the amount of carbon dioxide in the atmosphere has increased, so more carbon dioxide has dissolved in the oceans. This extra carbon dioxide makes seawater more acidic. This causes problems for some living organisms. Shellfish, for example, have difficulty making their shells.

## Global warming

Most scientists agree that the increasing concentration of carbon dioxide in our atmosphere causes global warming. An increase in the Earth's average air temperature will have far-reaching consequences:
- Weather patterns will change. There are likely to be more extreme weather events. Some areas will suffer from drought, whilst others will flood.
- Polar ice caps will melt. Sea levels will rise, and low-lying coastal areas will flood.

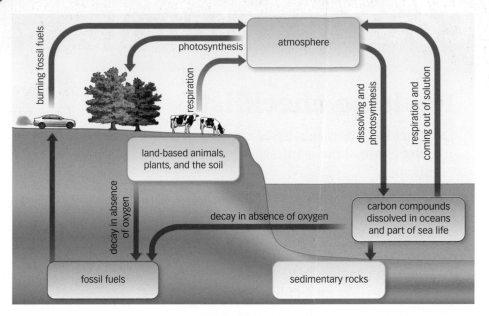

▲ The carbon cycle

▲ Deforestation, contributing to increased atmospheric carbon dioxide levels

### Questions

1 Name a natural process that adds carbon dioxide to the atmosphere.
2 Name a human activity that adds carbon dioxide to the atmosphere.

↓ E

3 Make a table to summarise the ways by which carbon dioxide enters and leaves the atmosphere.

↓ C

4 Describe and explain the problems caused by more carbon dioxide dissolving in the sea.

↓ A*

# Course catch-up

## Revision checklist

- Cracking using heat and a catalyst breaks down long-chain alkanes into smaller molecules, including alkenes.

- Alkenes are unsaturated hydrocarbons which contain C=C double bonds and have general formula $C_nH_{2n}$. Bromine water is used to test for C=C bonds.

- Alkenes are used to manufacture polymers. In polymerisation reactions, many small monomer molecules join to form one large polymer molecule.

- Polymers may have a wide range of properties. The uses of a polymer depend on its properties.

- The disposal of non-biodegradable polymers causes environmental problems.

- Ethanol is a molecule which can be formed from renewable sources (fermentation of sugars) or non-renewable sources (reaction of ethene + steam).

- Plant oils have high energy content, high boiling point, and are used as foods and in cooking.

- Plant oils can be hardened into margarine for use in foods by reacting with hydrogen (hydrogenation).

- Plant oils can be extracted from plant material by pressing, solvent extraction followed by distillation, or steam distillation.

- Plant oils do not dissolve in water, but can be made into emulsions by using emulsifiers.

- The Earth has a layered structure. The thin rocky crust floats on the mantle, and at the centre is the hot iron core.

- The crust is divided into tectonic plates. Convection currents in the mantle cause the plates to move. Earthquakes and volcanic eruptions occur at plate boundaries.

- The Earth's atmosphere has a stable composition (78% nitrogen, 21% oxygen + small amounts of other gases, including $CO_2$).

- The early atmosphere of the Earth was mostly $CO_2$ and water vapour plus small amounts of ammonia and methane.

- Processes which caused the atmosphere to change include rock formation, development of plant life (photosynthesis), dissolving of gases in the seas, and volcanic eruptions.

- The formation of complex molecules from the reaction of hydrocarbons with ammonia during lightning strikes may have helped life to form.

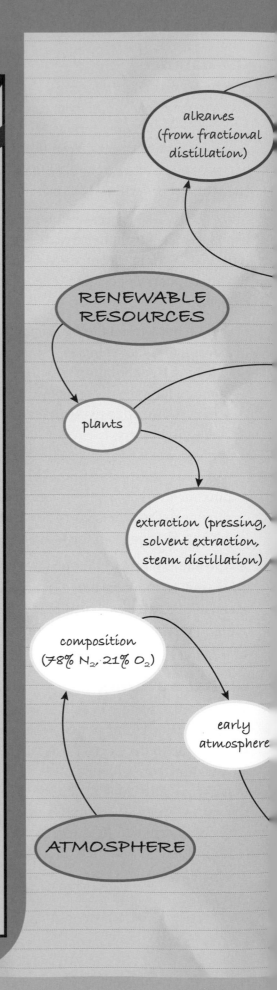

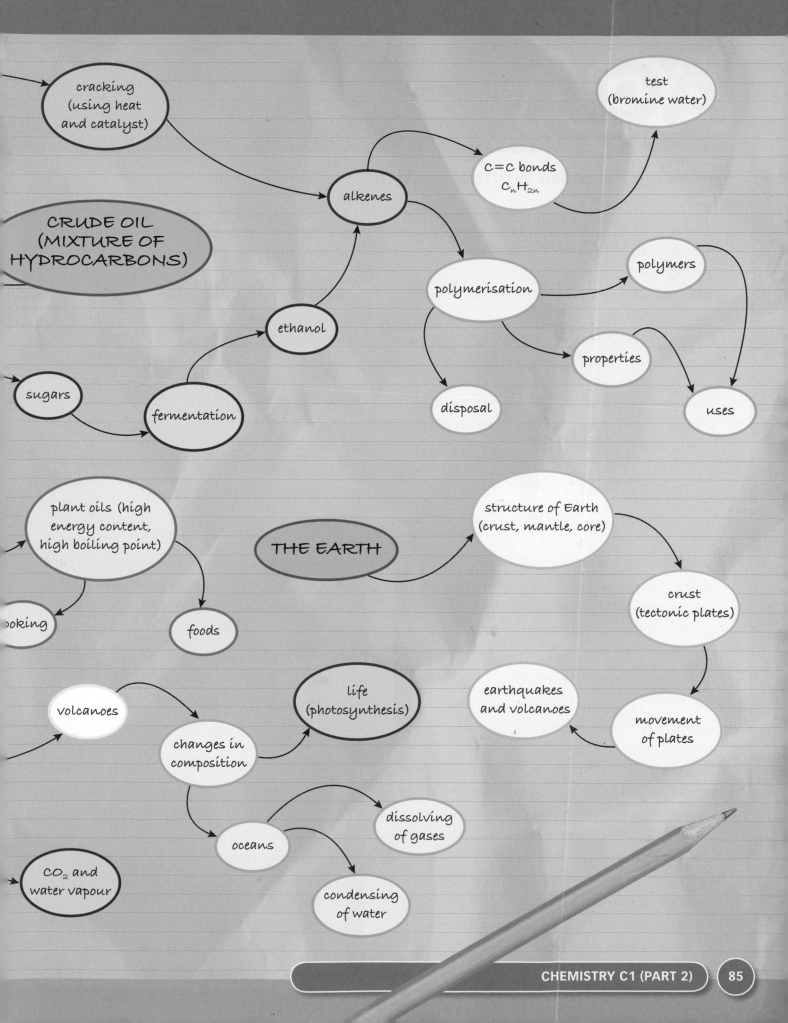

cracking
(using heat
and catalyst)

test
(bromine water)

C=C bonds
$C_nH_{2n}$

alkenes

CRUDE OIL
(MIXTURE OF
HYDROCARBONS)

polymers

polymerisation

ethanol

properties

sugars

fermentation

disposal

uses

plant oils (high
energy content,
high boiling point)

structure of Earth
(crust, mantle, core)

THE EARTH

crust
(tectonic plates)

ooking

foods

volcanoes

life
(photosynthesis)

earthquakes
and volcanoes

movement
of plates

changes in
composition

oceans

dissolving
of gases

$CO_2$ and
water vapour

condensing
of water

# Answering Extended Writing questions

In Brazil, ethanol has been used as a vehicle fuel for many years. Ethanol can be produced by the fermentation of plant sugars, such as those in sugar cane. It can also be produced from ethene gas. Ethene is made from crude oil.

Outline the advantages and disadvantages of producing ethanol from plant crops, compared to producing ethanol from ethene.

**The quality of written communication will be assessed in your answer to this question.**

---

Fermentasion is good because you can grow shugar cane every year. The ethene method needs lots of energy

**G–E**

Examiner: The candidate knows that ethanol made from sugar cane is a renewable resource, but has not used the scientific word to describe this advantage. The second point is also correct. The candidate has not made it obvious how the two processes compare. There are two spelling errors and one punctuation error.

---

Using ethanol made from sugar cane is carbon neutral. The sugar take in the same amount of carbon dioxide when they grow as the ethanol give out when it burn. Ethanol from ethene is not renewable. A disadvantage of ethanol from sugar is you should use land to grow food.

**D–C**

Examiner: The answer makes three correct points, and the spelling and punctuation are good. There are some grammatical errors. The answer explains the meaning of the term 'carbon neutral', but does not mention that making fertilisers for sugar cane crops causes carbon dioxide emissions.

---

Sugar cane is a renewable resource – you can grow it again. Ethene is a non-renewable resource, because it comes from crude oil. Fermentation happens at 37ºC. Ethene makes ethanol by reacting with steam at 300ºC. So fermentation needs lower energy inputs. But fermentation makes waste carbon dioxide. The other process makes no waste products. It is morally wrong to use land for fuel crops instead of food.

**B–A\***

Examiner: This answer clearly describes the advantages and disadvantages of the two processes. It is logically organised, and includes scientific terms that are used correctly. The spelling, punctuation, and grammar are accurate.
The answer would be even better if it made clear which statements refer to advantages and which to disadvantages.

# Exam-style questions

**1** Scientists now know that the Earth is made up of several layers:

core        crust        mantle

**A01** **a** Use the words above to label the different layers in this diagram.

**A01** **b** Which layer is:
    **i**   made of nickel and iron?
    **ii**  a source of important raw materials for the chemical industry?
    **iii** made up of a number of large tectonic plates?

**2** Large hydrocarbon molecules obtained from crude oil can be converted into smaller molecules by a process called cracking.

**A01** **a** What conditions are needed for a cracking reaction?

**A02** **b** Give one reason why cracking is important in the petrochemical industry.

**3** **a** The displayed formula of chloroethene is:

**A02**    **i**  Why is this molecule described as unsaturated?

**A02**    **ii** A student adds bromine water to a small sample of this substance. What would she observe?

    **iii** Complete this balanced equation to show the reaction which happens when the bromine is added:
    $C_2H_3Cl + Br_2 \rightarrow$

**b** Chloroethene is a monomer used in industry to manufacture the polymer poly(chloroethene).

**A02**    **i**  Draw out the displayed formula of poly(chloroethene).

**A02**    **ii** Poly(chloroethene) is used to make window frames. Suggest two properties which the polymer must have to make it suitable for this use.

## Extended Writing

**4** Olives grow in some parts of the world.
**A01** They are a good source of oil. This olive oil is used in cooking, is eaten as a food, and can be used as a fuel.
Write about why olives can be used in this way.

**5** Polymers are a very important type of
**A01** substance in today's society. However, many people are worried about how to dispose of them.
Explain why disposing of polymers is a particular problem.

**6** The atmosphere of the Earth today
**A02** contains about 21% oxygen, 78% nitrogen, and 1% other gases.
Describe how the early atmosphere may have been different to today's atmosphere.

G–E

D–C

B–A*

B–A*

G–E

D–C

B–A*

---

**A01** Recall the science

**A02** Apply your knowledge

**A03** Evaluate and analyse the evidence

# C2 Part 1

# Structures, properties, and uses

## Why study structures, properties, and uses?

Scientists have been uncovering the secrets of structures for centuries. We know what happens deep inside atoms. We know how particles give substances their properties. We know how to manipulate structures to make substances with perfect properties for particular purposes.

Today, scientists are excited about nanoscience – the study of structures measuring around one billionth of a metre. Will discoveries at the nanoscale lead to new cancer cures, greener energy technologies, and better computers?

In this unit you will learn how particles are arranged and joined together in metals, non-metals, and polymers, and about nanostructures. You will discover how the structures of substances influence their properties and uses. You will also learn about analysis, and how to calculate yields.

## You should remember

1  Everything is made up of tiny particles, called atoms.

2  Protons, neutrons, and electrons make up atoms.

3  Atoms can share electrons to form covalent bonds.

4  Atoms can join together by losing or gaining electrons to form ions, which form ionic bonds.

5  The properties and uses of substances depend on how their particles are arranged.

6  Chromatography separates the components of a mixture, and can help us to identify the components.

7  Chemical formulae give information about the types and numbers of atoms that make up a substance.

8  Word equations show the reactants and products involved in chemical reactions.

The picture is an electron micrograph of the end of a nanotube. Nanotubes consist of sheets of carbon atoms arranged in hexagons. The sheets are wrapped around each other to form a cylinder with a hollow core. Ten thousand nanotubes side by side would only be as wide as a human hair.

Nanotubes are 10 000 times stronger than steel, but have a much smaller density. Added to tennis racquets, they already give elite players more power and control. Filled with metals, they make the world's smallest bar magnets. In future, nanotubes might make miniature circuits, or ultra-lightweight planes. The possibilities are endless...

### Learning objectives

After studying this topic, you should be able to:

✔ explain how covalent bonds are formed

✔ draw dot and cross diagrams for simple molecules

### Key words

compound, molecular formula, molecule, covalent bond, dot and cross diagram

## Versatile gas

What do nitrogen fertiliser, explosives, household cleaner, and fish have in common?

These are all things that rely on ammonia. Ammonia is a raw material for making fertilisers and explosives. Ammonia solution is a household cleaner. Fish produce ammonia as a waste product. They excrete the ammonia into the water from their gills. Ammonia has a pungent smell – like toilets that need cleaning.

## Inside ammonia

Ammonia is a **compound**. A compound is a substance that is made up of two or more different elements, chemically combined. So in a compound, the elements are not just mixed up. Chemical bonds join the elements together.

The **molecular formula** of ammonia is $NH_3$. This shows that ammonia is made up of atoms of two elements – nitrogen and hydrogen. There are three atoms of hydrogen for every one atom of nitrogen.

A  The formula of water is $H_2O$. Name the two elements in this compound.

B  Write down the formulae that represent the compounds in this list: $H_2$ $Cl_2$ $CH_4$ $O_2$ $SiO_2$

90

# Joining atoms together

Ammonia gas exists as **molecules**. A molecule is a particle made up of two or more atoms chemically bonded together. In ammonia, each molecule consists of one atom of nitrogen joined to three atoms of hydrogen. The atoms are held together by **covalent bonds**. A covalent bond is a shared pair of electrons.

Covalent bonds form so that atoms can achieve stable electron arrangements. For example, a nitrogen atom has seven electrons. The electrons are arranged like this:

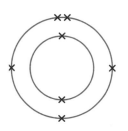

▲ Electronic structure of nitrogen    ▲ Electronic structure of hydrogen

There are five electrons in the highest occupied energy level, or outermost shell. This arrangement is not very stable. The nitrogen atom needs three more electrons in its highest energy level. It will then have the stable electronic structure of a noble gas, with eight electrons in its highest energy level.

A hydrogen atom has just one electron. It needs one more electron to achieve the stable electronic structure of the noble gas helium.

When nitrogen and hydrogen react to form ammonia, the two types of atom join together by *sharing* pairs of electrons. Each shared pair of electrons is one covalent bond. By sharing electrons, both the nitrogen and hydrogen atoms achieve stable electronic structures.

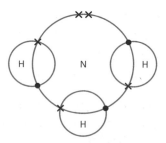

▲ In this **dot and cross diagram** for ammonia, crosses represent electrons from the nitrogen atom and dots represent electrons from the hydrogen atom. Each covalent bond is shown in a shared area as a dot and a cross.

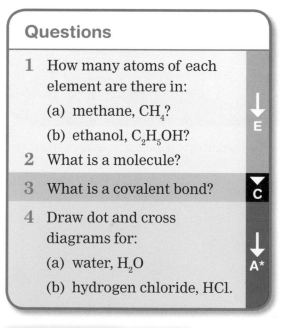

## Questions

1  How many atoms of each element are there in:

   (a) methane, $CH_4$?

   (b) ethanol, $C_2H_5OH$?

2  What is a molecule?

3  What is a covalent bond?

4  Draw dot and cross diagrams for:

   (a) water, $H_2O$

   (b) hydrogen chloride, HCl.

**Exam tip**

✓ When you draw dot and cross diagrams for covalent compounds, you only need to show the electrons in the highest occupied energy level (outermost shell).

## More molecules

Every day, rice crops release huge quantities of methane gas to the atmosphere. So do burping cows. Methane is a greenhouse gas. Its presence in the atmosphere contributes to global warming.

The central carbon atom in a methane molecule shares its outermost electrons with four hydrogen atoms. Each shared pair of electrons is a covalent bond. Here is a dot and cross diagram for methane:

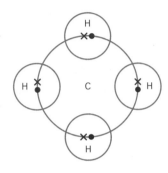

You can also represent the shared pairs of electrons like this:

In the diagram below, each line represents a shared pair of electrons. This is the **displayed formula** for methane. It shows the atoms in a molecule *and* the covalent bonds between them.

**Learning objectives**

After studying this topic, you should be able to:

✔ describe covalent bonding in simple molecules

✔ draw displayed formulae for simple molecules

▲ Cows release huge quantities of methane to the atmosphere

**Key words**

displayed formula, double covalent bond, giant covalent structure, macromolecule

**A** What is a displayed formula?

**B** Draw displayed formulae for ammonia, $NH_3$, and water, $H_2O$.

## Bonding in elements

There are covalent bonds in elements, too. The two atoms in a chlorine molecule, $Cl_2$, share a pair of electrons to make a single covalent bond.

The two oxygen atoms of an oxygen molecule share two pairs of electrons. This gives each oxygen atom the stable electronic structure of a noble gas, with eight electrons in the highest occupied energy level of each atom. The two shared pairs of electrons form a strong **double covalent bond**.

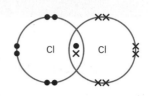

▲ This dot and cross diagram shows the electrons in the highest occupied energy levels of a chlorine molecule, $Cl_2$

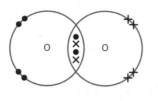

To find out about more about the properties of substances made from simple molecules, see spread C2.6.

## Macromolecules

Some covalently bonded substances are joined together in huge networks called **giant covalent structures**, or **macromolecules**.

Diamond is a form of the element carbon. In diamond, covalent bonds join each carbon atom to four other carbon atoms. The bonds are arranged like this:

▲ This dot and cross diagram shows the electrons in a hydrogen molecule, $H_2$

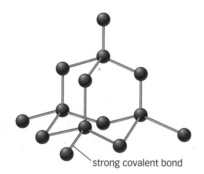

strong covalent bond

Silicon dioxide – the compound in most sand – also exists as macromolecules. Its atoms are joined together in a pattern like this:

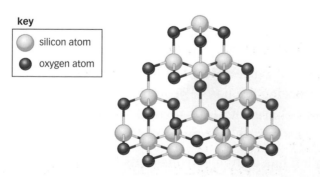

key
○ silicon atom
● oxygen atom

To find out about more about the properties of substances made from macromolecules, see spread C2.8.

**Exam tip** **AQA**

✔ There are covalent bonds in compounds of non-metals and in non-metal elements.

## Questions

1 What information is shown by a displayed formula?

2 Name two substances that have a giant covalent structure.

↓ E

3 Draw a displayed formula for a chlorine molecule.

↓ C

4 Explain why the two oxygen atoms in an oxygen molecule are joined by a double covalent bond.

↓ A*

5 How many covalent bonds does each silicon atom have in silicon dioxide?

## Learning objectives

After studying this topic, you should be able to:

✔ explain what ions are

✔ explain how positive and negative ions are formed by electron transfer

✔ use dot and cross diagrams to represent ions

✔ describe ionic bonding

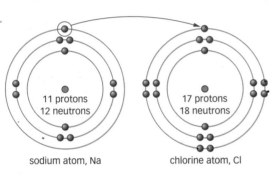

sodium atom, Na          chlorine atom, Cl

▲ When sodium and chlorine react together, each sodium atom transfers one electron to a chlorine atom

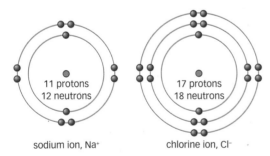

sodium ion, Na⁺          chlorine ion, Cl⁻

▲ The diagrams show the electron arrangements in sodium and chloride ions

**A** Explain how positively charged ions are formed.

**B** Neon is a noble gas. How many electrons are in its highest occupied energy level?

## Vital ion

Kerry runs a marathon. At the end of the race, she vomits. She is dizzy and confused. In hospital, the doctor tells Kerry she drank too much water. The concentration of sodium **ions** in her blood is too low. She needs an injection of sodium chloride solution to replace the lost sodium ions – now!

## Making ions

An ion is an electrically charged atom, or group of atoms. Ions form when atoms lose or gain electrons. Electrons are negatively charged, so:

- If an atom loses one or more electrons, it becomes a positively charged ion.
- If an atom gains one or more electrons, it becomes a negatively charged ion.

A sodium ion forms when a sodium atom transfers one of its electrons to an atom of a non-metal element, such as chlorine, in a chemical reaction:

- A sodium ion has 11 protons and 10 electrons. Overall, it has a charge of +1. Its formula is $Na^+$. You can also represent its electronic structure as $[2,8]^+$.
- A chloride ion has 17 protons and 18 electrons. Overall, it has a charge of –1. Its formula is $Cl^-$. You can also represent its electronic structure as $[2,8,8]^-$.

Each ion has eight electrons in its highest occupied energy level, like a noble gas. These electron arrangements are very stable.

# Ionic bonding

Compounds made up of ions are called **ionic compounds**. There are strong electrostatic forces of attraction between the oppositely charged ions. This is **ionic bonding**. Ionic bonds act in all directions and hold the ions in a regular pattern, called a **giant ionic lattice.** See spread C2.7 to find out about the properties of ionic compounds.

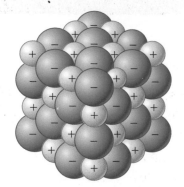

▲ The lattice structure of sodium chloride

## Did you know...?

Ions from the Sun stream towards the Earth and interact with its magnetic field, especially near the poles. Different substances in the upper atmosphere glow in different colours, giving a spectacular natural light show in the sky. This is known as the aurora borealis.

## Key words

ion, ionic compound, ionic bonding, giant ionic lattice

▲ The aurora borealis

## Questions

1 What is an ion?
2 What is an ionic compound?
3 Describe the bonding in an ionic compound.
4 Draw and annotate a diagram to explain how a sodium atom transfers electrons to a chlorine atom to make ions.
5 Explain how ionic and covalent bonding are different.

## Exam tip

- ✓ When a chlorine atom gains an electron, it forms a chloride ion, Cl⁻.
- ✓ Covalent bonding involves sharing electrons, but ionic bonding involves transferring electrons.

# 4: Making ionic compounds

## Learning objectives

After studying this topic, you should be able to:

- ✔ describe the reactions of the Group 1 metals with non metal elements
- ✔ represent the electronic structures of ions in ionic compounds

▲ In March 1930, Mahatma Gandhi led a non-violent protest against the British salt tax in colonial India

◀ Sodium reacts vigorously with chlorine to form sodium chloride

## White gold

Thousands of years ago, humans discovered that salt preserves food. Salt became valuable. The Romans built roads just to transport it, and salt taxes made rulers rich. Wars have been fought over what we think of as an everyday item!

## Making salt

Table salt is sodium chloride. It is made up of the elements sodium and chlorine. The salt we eat comes from the sea, or from underground rock salt. But you can make it in chemical reactions, too.

Flo's teacher heats a piece of sodium. He puts a gas jar of chlorine over it. There is a bright orange flame. Flo sees white clouds. Tiny white crystals of sodium chloride form on the sides of the gas jar.

Sodium is a metal. It is in Group 1 of the periodic table, the **alkali metals**. Each sodium atom has one electron in its highest occupied energy level, like all the other alkali metals. Chlorine is a non-metal. It is in Group 7 of the periodic table, the **halogens**. Each chlorine atom has seven electrons in its highest occupied energy level, like all the other halogens.

| 1 | 2 | | | | | | | | | | | 3 | 4 | 5 | 6 | 7 | 0 |
|---|---|---|---|---|---|---|---|---|---|---|---|---|---|---|---|---|---|
| | | | | | | H | | | | | | | | | | | He |
| Li | Be | | | | | | | | | | | B | C | N | O | F | Ne |
| Na | Mg | | | | | | | | | | | Al | Si | P | S | Cl | Ar |
| K | Ca | Sc | Ti | V | Cr | Mn | Fe | Co | Ni | Cu | Zn | Ga | Ge | As | Se | Br | Kr |
| Rb | Sr | Y | Zr | Nb | Mo | Tc | Ru | Rh | Pd | Ag | Cd | In | Sn | Sb | Te | I | Xe |
| Cs | Ba | La | Hf | Ta | W | Re | Os | Ir | Pt | Au | Hg | Tl | Pb | Bi | Po | At | Rn |
| Fr | Ra | Ac | | | | | | | | | | | | | | | |

alkali metals

the halogens

When sodium reacts with chlorine, each sodium atom transfers an electron to a chlorine atom. Two types of ions form – $Na^+$ and $Cl^-$. These ions make up the compound sodium chloride.

## Alkali metals and the halogens

All Group 1 elements react vigorously with Group 7 elements, particularly if the metal is heated first. Clouds of a **metal halide** are produced. Fluorine produces a metal fluoride, chlorine produces a metal chloride, bromine produces a metal bromide, and iodine produces a metal iodide. The metal halide produced depends on the metal used, too. For example:

lithium + bromine → lithium bromide

$$2Li + Br_2 \rightarrow 2LiBr$$

In all these reactions, each metal atom transfers one electron to a halogen atom. Two types of ion form:

- one with a single positive charge, such as $Li^+$
- one with a single negative charge, such as $Br^-$.

## Alkali metals and oxygen

Alkali metals react with other non-metals, too. For example, heating sodium in air makes sodium oxide.

sodium + oxygen → sodium oxide

$$4Na + O_2 \rightarrow 2Na_2O$$

Each oxygen atom has gained one electron from each of two sodium atoms. The sodium and oxide ions now have the stable electronic structures of noble gases. The formulae of the ions in sodium oxide are $Na^+$ and $O^{2-}$. It takes two sodium ions, $Na^+$, to balance the two charges on an **oxide ion**, $O^{2-}$, so the formula of sodium oxide is $Na_2O$.

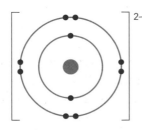

▲ You can represent the electronic structure of an oxide ion like this

## Learning objectives

After studying this topic, you should be able to:

✔ describe metallic bonding

▲ Lead is an excellent roofing material

**A** For a lump of lead, name the type of particle that is arranged in a regular pattern and the type of particle that is delocalised.

**B** Explain why electrons from the highest occupied energy level become delocalised in metals.

## Exam tip

✔ Be sure to mention delocalised electrons and positive ions when explaining metallic bonding.

## Rogues on the roof

It's midnight. Police are called to a Bristol church. Thieves are taking lead from the roof. But why? And what was lead doing on the roof in the first place?

Lead is a metal. It is bendy and waterproof, and an excellent roofing material. The thieves wanted to sell the lead as scrap, for £650 a tonne. The scrap lead might end up in car batteries, protecting underwater cables, or even on another church roof.

## Metallic bonding

Lead is a metal. Like all metals, it consists of a giant structure of atoms arranged in a regular pattern. Strong forces hold the structure together.

In metal atoms, the electrons in the highest occupied energy level are not strongly attracted to the nucleus. These electrons leave their atoms. They are **delocalised**, and free to move throughout the whole metal structure.

The atoms that have lost electrons are positive ions. They are arranged in a regular pattern to form a **giant metallic structure**. There are strong electrostatic forces of attraction between the positive ions and the moving delocalised electrons. This is **metallic bonding**.

To find out how metallic bonding influences the properties of metals, see spread C2.10.

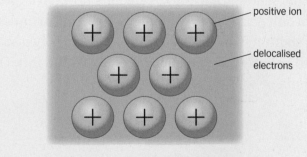
▲ Metallic structure

## How many electrons?

Potassium is in Group 1 of the periodic table. Its atoms have one electron in the highest occupied energy level. So potassium metal consists of a regular pattern of ions with a single positive charge, $K^+$. Delocalised electrons move throughout the metal. Attractive forces between the metal ions and the delocalised electrons hold the structure together.

Magnesium is in Group 2. Its atoms have two electrons in the highest occupied energy level. So magnesium metal contains $Mg^{2+}$ ions arranged in a regular pattern. There are two delocalised electrons for every magnesium ion.

### Key words

**delocalised**, **giant metallic structure**, **metallic bonding**

### Did you know...?

The Romans smelted about 80 000 tonnes of lead each year. They used it to make water pipes and added it to silver coins to make them less valuable.

▲ A Roman lead water pipe at the Roman Baths in Bath

### Questions

1 What is a delocalised electron?

2 Explain why metals contain positive ions.

↓ E

3 Describe metallic bonding.

▼ C

4 Use ideas about metallic bonding in sodium and magnesium to suggest why magnesium has a higher boiling point than sodium.

↓ A*

# 6: Molecules and properties

▲ Bromine is a volatile liquid at room temperature

## Bad smell

Guess which element is named from a Greek word meaning *to stink*?

The answer is bromine. Bromine is corrosive. Its vapours attack the eyes and lungs. Swallowing just 0.1 g of the liquid can be fatal. But bromine is not all bad. Its compounds are added to electronic goods and furniture to make them less likely to catch fire.

## Inside bromine

Bromine consists of simple molecules. A simple molecule is made up of just a few atoms joined together by strong covalent bonds. The non-metal elements oxygen and hydrogen consist of simple molecules. So do the compounds hydrogen chloride, methane, and ammonia.

## Properties of simple molecular substances

Substances that consist of simple molecules have low melting and boiling points compared to substances with ionic, metallic, or giant covalent structures.

| Substance | Melting point (°C) | Boiling point (°C) |
|---|---|---|
| chlorine | −101 | −35 |
| zinc chloride | 283 | 732 |
| nitrogen | −210 | −196 |
| diamond | 3550 | 4827 |
| water | 0 | 100 |
| sulfur dioxide | −73 | −10 |
| zinc | 420 | 907 |

**A** Look at the data in the table. Suggest which substances consist of simple molecules. Give reasons for your choices.

## Explaining low melting and boiling points

Bromine is one of just two elements that are liquid at room temperature. It consists of bromine molecules, $Br_2$. Strong covalent bonds hold the atoms together in each bromine molecule.

The forces between a molecule and its neighbours are much weaker. It is these **intermolecular forces**, not the covalent bonds, that must be overcome when bromine melts or boils.

This explains why bromine has low melting and boiling points. Other substances that consist of simple molecules have low melting and boiling points too.

| Substance | Boiling point (°C) |
|-----------|--------------------|
| fluorine | −118 |
| chlorine | −35 |
| bromine | 59 |
| iodine | 184 |

**B** Use the data in the table to describe the pattern in boiling points of the halogens.

**C** Explain why the halogens have low boiling points.

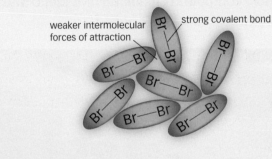

weaker intermolecular forces of attraction

strong covalent bond

▲ The diagram shows the covalent bonds and intermolecular forces in liquid bromine

### Did you know...?

Nitrogen gas becomes liquid at −196 °C. It is used to freeze blood and preserve genetic material, such as human eggs and sperm.

## Good conductors?

Substances that consist of simple molecules do not conduct electricity. This is because the molecules do not have an overall electrical charge.

### Questions

1 Describe three properties that are typical of substances that consist of simple molecules. ↓ E

2 Explain why methane does not conduct electricity.

3 Predict which has the higher boiling point – copper or methane. Explain your prediction. ↓ C

4 Use ideas about intermolecular forces to help you explain why ammonia has a low boiling point. ↓ A*

▲ Nitrogen gas condenses to become a liquid at −196 °C

# 7: Properties of ionic compounds

## Learning objectives

After studying this topic, you should be able to:

✔ explain how the properties of ionic compounds are linked to their structure

▲ The Taipei 101 building in Taiwan

## Did you know...?

Magnesium oxide does not only protect building materials from fire. It relieves heartburn and stomach ache, and is an effective laxative.

## Taipei 101

The Taipei 101 building in Taiwan is more than half a kilometre high. All the walls of its 101 storeys are covered with fire-resistant magnesium oxide. Magnesium oxide also protects the building's metal beams from fire.

## Inside ionic compounds

Magnesium oxide is an ionic compound, like sodium chloride. It has a repeating structure of positive magnesium ions and negative oxide ions, arranged in a giant ionic lattice. There are strong electrostatic forces of attraction between the oppositely charged ions. The forces act in all directions.

Magnesium oxide and sodium chloride are not the only ionic compounds. Most compounds made up of a metal and a non-metal are ionic.

▲ Crystals of copper chloride (×840)

## Explaining the properties of ionic compounds

Ionic compounds have very high melting points. A scientist heats up solid magnesium oxide to 714 °C. The heat energy disrupts the strong forces of attraction in the giant ionic lattice. The regular pattern of ions breaks down. The magnesium oxide melts and becomes liquid.

The scientist heats the liquid magnesium oxide even more, to 1422 °C. At this temperature, there is enough energy to overcome the strong forces of attraction between the ions. The magnesium oxide liquid boils and becomes a gas.

All ionic compounds have high melting points and high boiling points because of the large amounts of energy needed to break their many strong bonds.

# Electricity

Ionic compounds do not conduct electricity when they are solid. Their ions are not free to move from place to place to carry the current.

Ionic compounds conduct electricity when they are liquids, but their high melting points make this difficult to show in the laboratory. In ionic liquids, the ions are free to move from place to place to carry the current.

Ionic compounds also conduct electricity when they are dissolved in water. Again, the ions are free to move from place to place. Of course, not all ionic compounds are soluble in water.

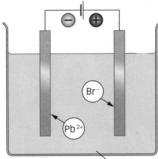

melted lead bromide

▲ Lead bromide is an ionic compound. Liquid lead bromide conducts electricity because its ions are free to move towards the electrodes.

**A** Describe the forces that hold the ions together in ionic compounds.

**B** Explain why magnesium oxide has a high boiling point.

**Exam tip**

✔ Solid ionic compounds do not conduct electricity. Liquid ionic compounds, and those dissolved in solution, do conduct electricity.

## Questions

1 Under what conditions does sodium chloride conduct electricity?

2 Name the type of force that holds the ions together in an ionic compound.

3 Suggest why magnesium oxide bricks are used to line the inside of furnaces.

4 Explain why ionic compounds have high melting points.

5 Look at the boiling points in the table. Suggest which compound is not an ionic compound. Give a reason for your decision.

| Name of compound | Boiling point (°C) |
|---|---|
| magnesium chloride | 1412 |
| lead bromide | 916 |
| sulfur dioxide | −10 |
| rubidium iodide | 1300 |
| silver chloride | 1550 |

## Learning objectives

After studying this topic, you should be able to:

✔ recognise the structures of diamond and graphite, and describe their properties

✔ explain some of the uses of diamond and graphite

✔ explain the properties of diamond and graphite in terms of their structure

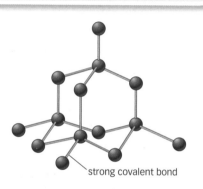

strong covalent bond

▲ Diamond structure

▲ Diamond's hardness and high melting point means it makes excellent dental drill bits

## Crazy carbon

**Diamond** and **graphite** are different forms of the same element – carbon. Diamond is lustrous, hard, and makes stunning jewellery. Graphite is soft and grey. Mixed with clay, it makes pencil 'leads'. How can these two forms of the same element be so different? The answer is in their structures.

## Diamond – the inside story

In diamond, each carbon atom forms strong covalent bonds with four other carbon atoms. The pattern is repeated to make a giant covalent structure.

▲ There are about 10 000 000 000 000 000 000 000 carbon atoms in this 0.2 g diamond

The strong covalent bonds explain two important properties of diamond:

• its extreme hardness
• its high melting point of 3550 °C.

Diamonds are transparent. They can be cut into shapes that allow light to pass through them so that they seem to sparkle. Their lustre and hardness mean that diamonds are valuable gemstone jewels.

Diamond does not conduct electricity because it has no charged particles that are free to move.

> A Explain why diamond is hard.
>
> B Give one use of diamond that depends on its hardness. What other property of diamond makes it suitable for this purpose?

# Graphite

The carbon atoms in graphite are arranged differently from those in diamond. Graphite has a layered structure. Each carbon atom is joined to three others by strong covalent bonds. But the forces *between* the layers are weak, so the layers can slide over each other easily. This is why graphite is slippery.

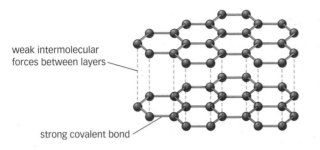

weak intermolecular forces between layers

strong covalent bond

▲ Graphite structure

Pencil 'lead' is not lead at all, but a mixture of graphite and clay. The black slippery graphite wears away on paper, leaving a black line.

▲ Pencil 'leads' are mainly graphite

Graphite's slipperiness also makes it useful in **lubricants**. At low temperatures, oil helps moving machine parts slide over each other easily. But oil breaks down at high temperatures. Graphite has a high melting point, and so makes a better high temperature lubricant than oil.

In graphite, one electron from each carbon atom is free to move. These delocalised electrons explain why graphite can conduct electricity. This property means that graphite is useful as **electrodes** in electrolysis. The delocalised electrons also allow graphite to conduct heat. The ability to conduct heat and electricity makes graphite like metals.

## Questions

1 Name the element of which graphite and diamond are different types.

2 Make a table to compare the properties of diamond and graphite.

3 Explain why diamond is used in drills, cutting tools, and jewellery.

4 Explain why graphite is used in pencils, lubricants, and electrodes.

5 In terms of their structure, explain why:
   (a) Graphite conducts electricity but diamond does not.
   (b) Graphite is slippery but diamond is hard.
   (c) Graphite and diamond both have high melting points.

E

C

A*

## Learning objectives

After studying this topic, you should be able to:

- ✔ explain what fullerenes are, and why they are useful
- ✔ explain the meanings of the terms nanoscience and nanoparticles
- ✔ describe some applications of nanoscience

## Key words

**fullerene**, **buckminsterfullerene**, nanoparticle, nanoscience, nanotube

## Did you know...?

Three scientists – Robert Curl, Harold Kroto, and Richard Smalley, discovered buckminsterfullerene in 1985. It was named after an American architect who designed 'geodesic domes'. These have the same sort of arrangement of hexagons and pentagons as buckminsterfullerene.

The Eden project in Cornwall has giant geodesic domes

## Fabulous fullerenes

Diamond and graphite are not the only forms of carbon. The element can also exist as **fullerenes**. Fullerenes are a type of carbon based on hexagonal rings of carbon atoms. Their properties are amazing!

The most common fullerene is **buckminsterfullerene**. It consists of molecules containing 60 carbon atoms joined together to form a hollow sphere. Its chemical formula is $C_{60}$.

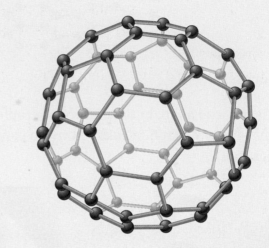

Fullerenes are hollow. The space inside is big enough for atoms and small molecules to fit in. Scientists have discovered how to 'cage' radioactive metal atoms and drug molecules inside fullerenes. These fullerenes can be used to deliver drugs into the body. For example, they can be coated with chemicals that make them gather next to cancer cells after being injected into the body. In this way, cancer drugs can be delivered to their targets without damaging normal cells.

**A** What are fullerenes?

**B** Explain why fullerenes can be used to deliver drugs into the body.

# Nanoscience

Fullerenes are examples of **nanoparticles**. Nanoparticles are tiny particles made up of a few hundred atoms. They measure between about 1 nanometre and 100 nanometres across. A nanometre is one billionth of a metre, so nanoparticles are much too small to see with a microscope. **Nanoscience** is the study of nanoparticles.

Nanoparticles are proving to be incredibly useful, and scientists keep on thinking of new ways of using their unique properties. For example:

- Fullerene particles join together to make **nanotubes**. These have a huge surface area compared to their volume. So they make excellent catalysts, to speed up reactions.
- Nanotubes are the strongest and stiffest materials ever discovered. So they are useful for reinforcing graphite tennis racquets.

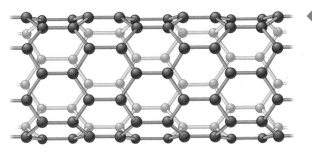

◀ The structure of a nanotube

The properties of nanoparticles are different from those of the same substances in normal-sized pieces. For example, titanium dioxide is a white solid. It is used in house paint. But titanium dioxide nanoparticles are so small they cannot reflect visible light. You cannot see them. This means they make excellent sun block creams. They protect your skin from harmful ultraviolet light without making your skin look white.

The amazing properties of nanoparticles could lead to many exciting developments in technology, including the development of

- new types of computer
- new coatings (in self-cleaning ovens and windows, for example)
- sensors that detect substances in tiny amounts
- very strong, light building materials.

▲ This water droplet is resting on a water-repellent surface. The surface contains nanoparticles that increase the contact angle between it and the water. These surfaces can withstand corrosion and stay clean.

## Questions

1. Explain the meanings of the words nanoparticle and nanoscience.
2. Describe three uses of nanoparticles.
3. Explain why nanotubes make good catalysts.
4. Explain why nanoparticles are added to the materials used to make some tennis racquets.
5. Describe some social and economic benefits of nanoscience.

↓ E

↓ C

↓ A*

## Learning objectives

After studying this topic, you should be able to:

✔ explain how the properties of metals are linked to their structure

✔ explain why alloys have different physical properties to the elements from which they are made

✔ describe the properties and uses of shape memory alloys

▲ Tanya has gold nails!

**A** Tanya sees an advert for getting her mobile phone plated with 24 carat gold, which is 99.9% gold. Suggest why 18 carat gold might be more suitable for this purpose.

**B** Use ideas about structure to explain why gilding alloy (95% copper and 5% zinc) is harder than pure copper.

## Clever copper

Electricians choose copper for their cables. Copper is bendy, and a good conductor of electricity.

◀ Copper cables

Metallic structure explains the properties of copper and other metals. There is a diagram showing metallic structure on spread C2.5. The positively charged metal ions are packed tightly in layers. The layers slide over each other easily. This explains why metals can be bent and shaped.

> Metals conduct electricity because their delocalised electrons are free to move throughout the metal to carry the current.

## Bling bling

Tanya buys gold fingernails to stick over her own nails. They are plated with 18 carat gold. Eighteen carat gold is not pure gold. Pure gold is too soft for false fingernails. It scratches easily and wears away quickly when it rubs against harder materials.

Eighteen carat gold is an **alloy**. An alloy is a mixture of a metal with one or more other elements. The physical properties of an alloy are different from the properties of the elements in it. Eighteen carat gold is a mixture of 75% gold mixed with copper and silver. The copper and silver atoms distort the layers in the gold structure, making it more difficult for them to slide over each other. This makes gold alloys harder than pure gold.

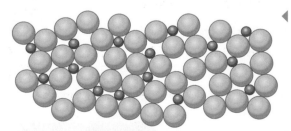

◀ An alloy is a mixture of a metal with small amounts of one or more other elements

# Smart alloys

Do you wear dental braces? If so, you have experienced **shape memory alloys** in action. Shape memory alloys – also called smart alloys – remember their original shapes. If a smart alloy is bent or twisted it keeps its new shape. But heating a smart alloy above a certain temperature makes the alloy return to its original shape.

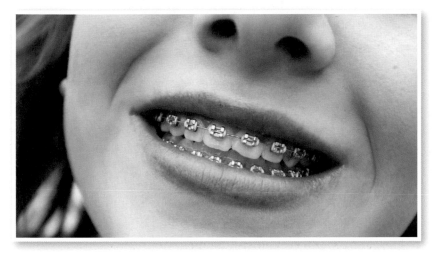

▲ Dental braces

Some of the best smart alloys are mixtures of nickel and titanium, called nitinol. Dental braces are made of nitinol wire. The wire exerts a constant force on the teeth to return the teeth to their correct positions.

Smart alloys have one main problem – they break easily if they are bent or twisted too much.

## Key words

alloy, shape memory alloy

## Did you know...?

Some glasses frames are made of nitinol. If you sit on them, you can gently warm the frame to return them to their original shape.

## Questions

1  Name the bonds found in metals.

2  Explain why electrical wires are made from copper.

↓ E

3  Describe two uses of smart alloys. Explain how their properties make them suitable for these purposes.

4  Explain why alloys are often harder than the metals that are mixed to make them.

↓ C

5  Describe metallic bonding, and explain why metals are easy to bend and how they conduct electricity.

↓ A*

## Exam tip

✔ In the exam, you need to be able to explain why a metal is suitable for a particular use. Sometimes, but not always, you may be given data to analyse.

## Two sorts of poly(ethene)

The bag and the bottle below are both made from poly(ethene). So why are their properties so different?

There are two types of poly(ethene). The bag is made from **low density poly(ethene)**, LDPE. The bottle is made from **high density poly(ethene)**, HDPE.

Each type of poly(ethene) has its own properties.

|  | LDPE | HDPE |
|---|---|---|
| density (g/cm$^3$) | 0.92 | 0.95 |
| strength (MPa) | 12 | 31 |
| transparency | good transparency | less transparent |
| relative flexibility | flexible | stiff |

The two types of poly(ethene) are both made from the same starting monomer – ethene. But they are made under different conditions.

|  | LDPE | HDPE |
|---|---|---|
| temperature (°C) | 100–300 | 300 |
| pressure (atm) | 1500–3000 | 1 |
| catalyst or initiator | oxygen or peroxide initiator | aluminium-based metal oxide catalyst |

The properties of all polymers depend on what they are made from and the conditions under which they are made.

A Name two types of poly(ethene).

B Use the data in the first table to explain why it is better to make bottles from HDPE than LDPE.

# Thermoplastic and thermosetting polymers

Some plastics, such as poly(ethene), soften easily when they are warmed. It is easy to mould them into new shapes, so they can be recycled. These are **thermosoftening polymers**. They consist of individual, tangled polymer chains.

> The forces of attraction between the separate polymer chains are weak.

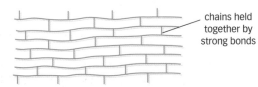

weak forces between the separate polymer chains

▲ Polymer chains in a thermosoftening polymer

Some plastics cannot be recycled because they do not melt when they are heated. These are **thermosetting polymers**. Thermosetting polymers consists of polymer chains with cross-links between.

> These cross-links are strong intermolecular bonds.

chains held together by strong bonds

▲ Polymer chains in a thermosetting polymer

## Questions

1 Describe three differences in the properties of LDPE and HDPE.

2 What is a thermosetting polymer?

3 Suggest some uses for LDPE and HDPE.

4 Explain why it is possible to melt thermosoftening polymers, but not thermosetting ones.

5 Draw a table to show the differences between thermosetting and thermosoftening polymers, and the reasons for these differences.

↓ E

▼ C

↓ A*

## Exam tip

✔ Thermosoftening plastics melt because they consist of individual chains. Thermosetting plastics have cross-links, so cannot melt.

## Learning objectives

After studying this topic, you should be able to:

- ✔ work out the mass number and atomic number of an atom
- ✔ explain what an isotope is

## Key words

sub-atomic particle, nucleus, proton, neutron, electron, atomic number, mass number, isotope

## Brain scan

Scans like this have unlocked some of the brain's deepest secrets. It is thanks to an understanding of atoms – and what goes on inside them – that scientists have been able to develop techniques like this.

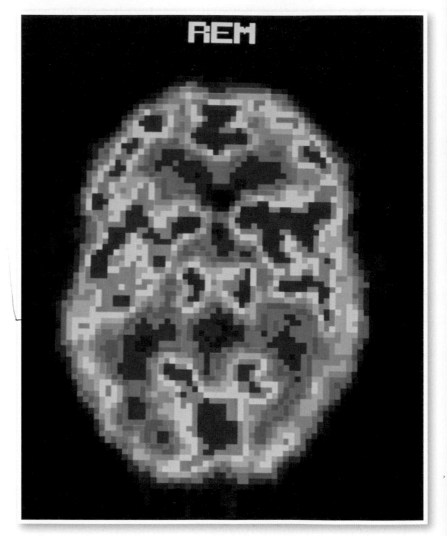

▲ A scan of a human brain

## Inside atoms

Atoms are the smallest part of an element that can exist. The diameter of a typical atom is 0.000 000 01 cm. But it is made up of even tinier particles, called **sub-atomic particles**.

In the centre of an atom is the **nucleus**. The nucleus is made up of **protons** and **neutrons**. The nucleus is surrounded by **electrons**. The table shows the masses and charges of these sub-atomic particles.

| Name of particle | Relative mass | Relative charge |
|---|---|---|
| proton | 1 | +1 |
| neutron | 1 | 0 |
| electron | very small | −1 |

## Identifying atoms

If you look for sodium on the periodic table, you will see it is represented like this:

$$^{23}_{11}\text{Na}$$

The number 11 is the **atomic number** of sodium. It is equal to the number of protons in a sodium atom, and also the number of electrons. The number 23 is the **mass number**. It is equal to the total number of protons and neutrons in a sodium atom. The number of neutrons is the mass number minus the atomic number. For example $^{23}_{11}\text{Na}$ contains 12 neutrons (23 – 11).

**Exam tip**

✔ You will be given a copy of the periodic table in the exam.

> **A** Work out the number of protons, neutrons, and electrons in an atom of $^{12}_{6}\text{C}$.
>
> **B** Work out the number of protons, neutrons, and electrons in an atom of $^{197}_{79}\text{Au}$.

## Isotopes

Atoms of the same element can have different numbers of neutrons. This means they have different mass numbers. Atoms of the same element which have different numbers of neutrons are called **isotopes**.

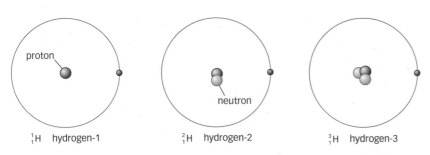

| $^{1}_{1}\text{H}$ hydrogen-1 | $^{2}_{1}\text{H}$ hydrogen-2 | $^{3}_{1}\text{H}$ hydrogen-3 |

▲ Three different isotopes of hydrogen. They contain the same number of protons and electrons, but different numbers of neutrons.

Every chlorine atom contains 17 protons and 17 electrons. About 75% of chlorine atoms have 18 neutrons. The other 25% of chlorine atoms have 20 neutrons. So there are two isotopes of chlorine. One has a mass number of 35 (17 + 18) and the other has a mass number of 37 (17 + 20).

### Questions

1 What is the charge on a proton?

2 Give the relative mass of a proton and the relative mass of an electron.

3 Define the term 'atomic number'.

4 What is an isotope?

5 Write a table to show the numbers of protons, neutrons, and electrons in the seven naturally occurring isotopes of mercury: $^{202}_{80}\text{Hg}$ $^{200}_{80}\text{Hg}$ $^{199}_{80}\text{Hg}$ $^{201}_{80}\text{Hg}$ $^{198}_{80}\text{Hg}$ $^{204}_{80}\text{Hg}$ $^{196}_{80}\text{Hg}$.

## Learning objectives

After studying this topic, you should be able to:

✔ calculate the relative formula mass of a given substance

## Did you know...?

A gold atom has a mass of $3 \times 10^{-21}$ kg. That's just 0.000 000 000 000 000 000 003 kg.

▲ Atoms of gold piled on top of a layer of carbon atoms

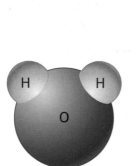

▲ The $M_r$ of water is 18. Each water molecule has 1.5 times the mass of a carbon atom.

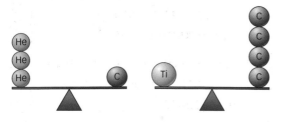

▲ Three helium atoms have the same mass as one carbon atom, and one titanium atom has the same mass as four carbon atoms

## Relative atomic mass

Individual atoms are tiny. They have very little mass. The gold atoms in the photograph are less than a millionth of a millimetre in diameter. The atoms have such small masses that it is more useful to use their **relative atomic mass** rather than their actual mass in kilograms.

Carbon atoms are the standard atom against which all the others are compared. The relative atomic mass, $A_r$, of the most common carbon isotope is exactly 12. Atoms with an $A_r$ of less than 12 have less mass than a carbon atom, and those with an $A_r$ greater than 12 have more mass than a carbon atom.

The relative atomic mass of an element is an average value for the isotopes of the element, taking into account their relative amounts. For example, about 75% of chlorine atoms have a mass number of 35. The other 25% have a mass number of 37. The relative atomic mass of chlorine is 35.5, an average of the masses of the two isotopes, taking into account their relative amounts.

> **A** How many helium atoms have a total mass equal to the mass of one titanium atom?

# Relative formula mass and moles

The chemical **formula** of a substance tells you the number of each type of atom in a unit of that substance. For example, the formula for water is $H_2O$. It shows that each water molecule consists of two hydrogen atoms and one oxygen atom, joined together.

The **relative formula mass** or $M_r$ of a substance is the mass of a unit of that substance compared to the mass of one carbon atom. It is worked out by adding together all the $A_r$ values for the atoms in the formula.

Scientists say that the relative formula mass of a substance, in grams, is one **mole** of that substance. So the mass of one mole of carbon atoms is 12 g.

**Key words**

relative atomic mass, formula, relative formula mass, mole,

## Worked example 1

What is the $M_r$ of water, $H_2O$?
$A_r$ values: H = 1, O = 16
$M_r$ of $H_2O$ = 1 + 1 + 16 = 18

## Worked example 2

What is the $M_r$ of magnesium hydroxide, $Mg(OH)_2$?
$A_r$ values: Mg = 24, O = 16, H = 1
$M_r$ of $Mg(OH)_2$ = 24 + [2 × (16 + 1)]
          = 24 + 34 = 58
(Notice that the 2 outside the brackets in $Mg(OH)_2$ means that there are two oxygen atoms and two hydrogen atoms.)

## Worked example 3

What is the mass of one mole of magnesium hydroxide?
$M_r$ of $Mg(OH)_2$ = 58, so the mass of one mole = 58 g

**B** What is the relative formula mass of magnesium oxide, MgO?

## Questions

Use the periodic table to find the answers to Questions 1 and 2.

1 What are the relative atomic masses of nitrogen, chlorine, and sodium?

2 What is the relative formula mass of:
   (a) oxygen, $O_2$
   (b) carbon dioxide, $CO_2$
   (c) ammonia, $NH_3$
   (d) sodium chloride, NaCl.

3 What is the relative formula mass of aluminium hydroxide, $Al(OH)_3$?

4 What is the relative formula mass of ammonium sulfate, $(NH_4)_2SO_4$?

E

C

A*

## Learning objectives

After studying this topic, you should be able to:

✓ describe how to use paper chromatography to identify food additives

✓ explain how gas chromatography separates the substances in a mixture

## Identifying food colourings

Tamara buys some sweets. She wants to know what colourings they have in them. She sets up a **paper chromatography** investigation to separate the colourings. She obtains this **chromatogram**:

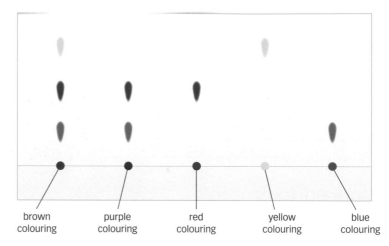

brown colouring | purple colouring | red colouring | yellow colouring | blue colouring

▲ The chromatogram shows that the brown colouring is a mixture of blue, red, and yellow dyes

In chromatography, a **mobile phase** moves through a **stationary phase**. Tamara added the sweet colouring sample to the stationary phase (paper). The mobile phase (in this case water) carried the chemicals in the sample through the stationary phase. Each compound moved at a different speed. So, the compounds in the mixture separated.

> **A** Which two food colourings are mixed in the purple dye?

## Drunk driver?

A police officer stopped Darren, who was driving his lorry erratically. She thought Darren might be drunk. Darren went to the police station but the breath test machine was broken. So he gave a urine sample instead.

The police officer sent the urine sample to a forensic laboratory. At the laboratory, scientists used **gas chromatography** to measure the exact concentration of alcohol in the urine.

# Gas chromatography

Gas chromatography is an instrumental method for detecting, identifying, and measuring chemical compounds.

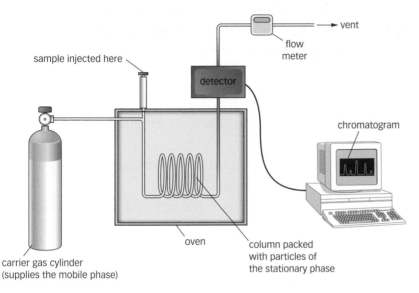

▲ Inside a gas chromatography instrument

Here's how it works:

- The urine sample is heated so that it becomes a mixture of vapours.
- A carrier gas, usually helium, mixes with the vapours. The gas is the mobile phase.
- The carrier gas takes the mixture of different vapours from the urine sample through a column packed with a solid material. The solid material in the column is the stationary phase.
- Different substances in the vapour mixture travel through the column, towards the detector, at different speeds. They become separated.

But was Darren over the limit? See spread 2.15 to find out how data from the gas chromatography analysis will provide the answer.

See spread 2.15

## Key words

paper chromatography, chromatogram, mobile phase, stationary phase, gas chromatography

## Did you know...?

The word 'chromatography' comes from the Greek word for colour.

## Questions

1 What is the stationary phase in paper chromatography?

2 In paper chromatography, what happens to the substances in the mixture?

3 What is the purpose of gas chromatography?

4 Name the commonly used carrier gas in gas chromatography.

5 Describe the stages by which the compounds in a sample are separated in gas chromatography.

↓E  ↓C  ↓A*

▲ Darren went to court and lost his licence and his job

## Interpreting gas chromatograms

Was Darren over the limit? Read on to discover the answer …

The gas chromatography column separates the substances in Darren's urine as they move towards the detector at the end of the column:

• The substance that travels most quickly reaches the detector first. This substance has the shortest **retention time**.
• The detector sends a signal to a recorder, which draws a peak on a chromatogram.
• The other substances reach the detector, one by one. The one that arrives last has the longest retention time.
• Scientists look at the chromatogram. Each peak represents one of the substances in the original mixture.
• The number of peaks shows the number of compounds present in the original mixture.
• The time taken for a substance to travel through the column helps to identify the substance.
• The relative areas under the peaks show the relative amounts of each of the compounds in the mixture.

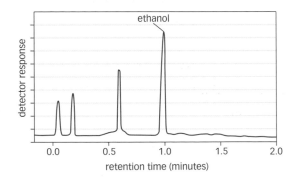

▲ A simplified version of part of the chromatogram from Darren's urine. It shows that there is ethanol (alcohol) in his urine. The scientists interpreted the data and said he was over the legal limit.

**A** A gas chromatogram has three peaks. What does this tell us about the mixture being analysed?

**B** Why is it useful to know the time taken for a substance to travel through a gas chromatography column?

## Mass spectrometry

Often, the detector in gas chromatography is a **mass spectrometer**. The mass spectrometer identifies the separated substances from gas chromatography very quickly and accurately, and can detect very small quantities.

The mass spectrometer can also give the relative molecular mass of each of the substances separated in the column.

### Mass spectrograph

Mass spectrometers produce mass spectrographs which look something like this.

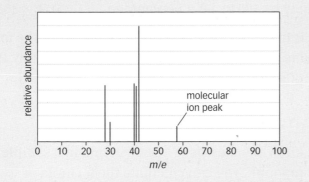

The peak on the right is the **molecular ion peak**. The mass of the molecular ion represents the relative formula mass of the molecule.

## Advantages of instrumental methods

Like all instrumental methods for detecting and identifying elements and compounds, gas chromatography and mass spectrometry are

- accurate
- sensitive
- quick.

▲ Researcher analysing results from a mass spectrometer

### Exam tip

✔ Gas chromatography separates substances. Mass spectrometry identifies them.

### Questions

1 On a gas chromatogram, what does the number of peaks show?

2 Which substance travels more quickly through a gas chromatography column – one with a shorter or one with a longer retention time?

3 Explain how a gas chromatogram indicates the relative amounts of the substances in a mixture.

4 Give three benefits of using instrumental methods to analyse mixtures.

5 Explain the purpose of using a mass spectrometer as the detector in gas chromatography.

## Learning objectives

After studying this topic, you should be able to:

- ✔ calculate the percentage of an element in a compound
- ✔ calculate empirical formulae
- ✔ calculate masses from equations

**A** Calculate the percentage by mass of potassium in potassium hydroxide, KOH.

**B** Calculate the percentage by mass of nitrogen in ammonium nitrate, $NH_4NO_3$.

▲ Many farmers add fertilisers to crops to improve their yields

## Calculating the percentage by mass of an element in a compound

Plants need potassium ions. So potassium chloride fertilisers improve crop yields. You can calculate the **percentage by mass** of potassium in the compound like this:

- Write down the formula of the compound.    KCl
- Use $A_r$ values to calculate the relative formula mass.    $39 + 35.5 = 74.5$
- Divide the $A_r$ of potassium by the $M_r$ of KCl.    $39 \div 74.5 = 0.52$
- Multiply by 100 to get a percentage.    $0.52 \times 100 = \textbf{52\%}$

If there is more than one atom of an element in the formula, you need to multiply the $A_r$ value of that element by the number of atoms shown.

## Calculating empirical formulae

If you know the masses of the elements in a sample of a compound, you can calculate the formula of the compound.

### Worked example 1

A sample of a compound contains 1.2 g of carbon and 3.2 g of oxygen. What is its formula?

| | carbon | oxygen |
|---|---|---|
| Mass of each element | 1.2 g | 3.2 g |
| $A_r$ (from periodic table) | 12 | 16 |
| mass divided by $A_r$ | 0.1 | 0.2 |
| simplest ratio | 1 | 2 |

So the formula of the compound is $CO_2$.

## Calculating masses of reactants and products from equations

In a chemical reaction, atoms are not created or destroyed. So the total mass of reactants is equal to the total mass of products.

**Key words**

percentage by mass

> ## Worked example 2
>
> Catherine heats 200 g of calcium carbonate. It decomposes to make calcium oxide and carbon dioxide. How much calcium oxide does she make?
>
> $$CaCO_3 \rightarrow CaO + CO_2$$
>
> $M_r$ of $CaCO_3$ = 40 + 12 + (16 × 3) = 100
>
> = 40 + 16 = 56
>
> Use ratios to work out the answer:
>
> 100 g of calcium carbonate would make 56 g of calcium oxide, so
>
> 200 g of calcium carbonate makes 112 g of calcium oxide.

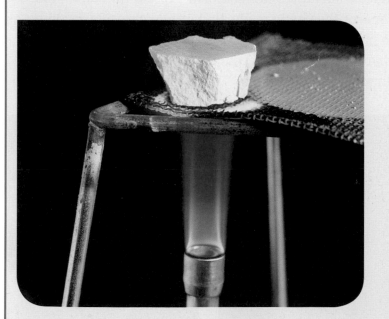

▲ Heating calcium carbonate

## Questions

1 Calculate the percentage by mass of carbon in carbon dioxide, $CO_2$. ↓ E

2 A sample of a compound contains 3.9 g of potassium, 1.2 g of carbon, and 1.4 g of nitrogen. What is its formula?

3 A compound is made up of 27% sodium by mass, 16% nitrogen by mass, and 56% oxygen by mass. What is its formula? ↓ A*

4 Edward heats 24 g of carbon in air. What mass of carbon dioxide does he make?

## Less than 100%

In most chemical reactions, chemists want to make as much product as possible. You can use reaction equations to calculate the maximum mass of a product you could expect to make (see spread C2.16). But, even though atoms are not created or destroyed in chemical reactions, you are likely to make a smaller mass than the calculated mass. This might be because:

- Some of the product was lost when you separated it from the reaction mixture. For example, if you are separating a solid product from a solution, some of the solid product might get stuck in the filter paper.

- The reactants may react in ways that are different from the expected reaction. For example, if you burn lithium in air to make lithium oxide, you might also make lithium nitride.

- The reaction is reversible. This means that the products of the reaction can react to make the original reactants. For example, ammonia and hydrogen chloride react to make solid ammonium chloride. But at the same time, some of the newly-made ammonium chloride decomposes to make ammonia and hydrogen chloride again. You can represent **reversible reactions** like this:

ammonium chloride $\rightleftharpoons$ ammonia + hydrogen chloride

▲ Some product may be lost during filtration

**A** Give three possible reasons for the actual yield in a reaction being less than the maximum theoretical yield.

**B** Suggest an economic reason for chemists in a chemical company wishing to make as much product as possible.

▲ Burning lithium in air to make lithium oxide and lithium nitride

▲ The reaction of ammonia and hydrogen chloride is reversible

# Yield

The **yield** of a substance is how much there is of it after a chemical reaction. Usually, there is a difference between the **actual yield** and the **maximum theoretical yield**:

- The maximum theoretical yield is the expected mass calculated from the reaction equation, using the masses of substances that react.
- The actual yield is the mass of product made.

## Percentage yield

The **percentage yield** compares these amounts. Use this formula to calculate it:

$$\text{percentage yield} = \frac{\text{actual yield}}{\text{maximum theoretical yield}} \times 100$$

Mr Merry heated a piece of sodium metal in chlorine gas. He calculated a maximum theoretical yield for sodium chloride of 10 g. But the actual yield was only 8 g. This means that the percentage yield was 80%:

$$\text{percentage yield} = \frac{8}{10} \times 100 = 80\%$$

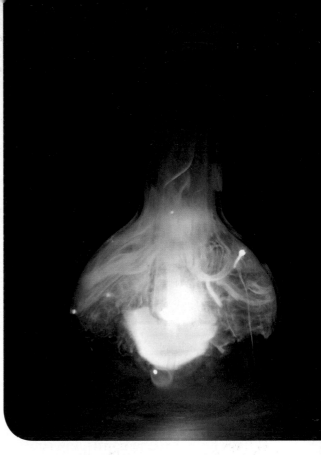

▲ The actual yield of most reactions is less than the maximum theoretical yield

## Questions

1. What is a reversible reaction?

2. Sarah heats magnesium in air to make magnesium oxide. Suggest why the actual yield might be less than the maximum theoretical yield she calculated before starting.

3. Zachary mixes lead nitrate and potassium iodide solutions to make solid lead iodide. He filters the mixture to separate the solid product from the solution. Suggest why the actual yield might be less than the maximum theoretical yield he calculated before starting.

4. Grayson burns a sample of magnesium in air. He calculates that the maximum theoretical yield of magnesium oxide is 5 g, but ends up with only 4.5 g. What is the percentage yield?

**Exam tip**

- If you are asked to calculate the percentage yield in the exam, you will be given the actual yield and the theoretical yield.

### Key words

reversible reaction, yield, actual yield, maximum theoretical yield, **percentage yield**

# Course catch-up

## Revision checklist

- In covalent bonding atoms share electron pairs, forming molecules or giant covalent structures.
- In ionic bonding atoms transfer electrons, forming ions. Positively and negatively-charged ions attract each other in giant ionic lattices.
- Ionic compounds form when metals react with non-metals.
- Dot and cross diagrams show the arrangement of electrons in ionic and covalent substances.
- Ions in ionic compounds and atoms in covalent compounds have stable electronic structures.
- In metallic bonding, layers of closely-packed ions are surrounded by a sea of delocalised electrons.
- Substances made up of simple molecules have low melting and boiling points and do not conduct electricity. Ionic compounds have high melting points and conduct electricity when molten or in solution.
- Diamond and graphite are forms of carbon with giant covalent structures. Fullerenes are another form of carbon used to form nanotubes.
- Metals conduct electricity because of delocalised electrons.
- The properties of metals are altered by forming alloys.
- Thermosetting polymers have cross-links between chains (unlike thermosoftening polymers). They cannot be melted.
- The nucleus of an atom contains protons (positive) and neutrons (uncharged).
- Atomic number is the number of protons in the nucleus of an atom. Mass number is the total number of protons and neutrons in the nucleus of an atom.
- Isotopes have the same atomic number but different mass numbers.
- The relative formula mass of a substance, in grams, is called one mole.
- Gas chromatography and mass spectrometry are instrumental techniques used to separate and identify compounds.
- Empirical formulas show the simplest ratios of the elements in a compound.
- Percentage yields can be low because product is lost in separation, or the reaction is reversible.

ATOMIC STRUCTURE

nucleu

isotopes

same atomic num
but different ma
number

INSTRUMENTAL METHODS

mass spectrometry

gas chromatograph

mass o
one mol

relative molecular mass

reacting mass

CALCULATIONS

empiri
formu

percentage yields

losses due to
reversible reactic
separation etc

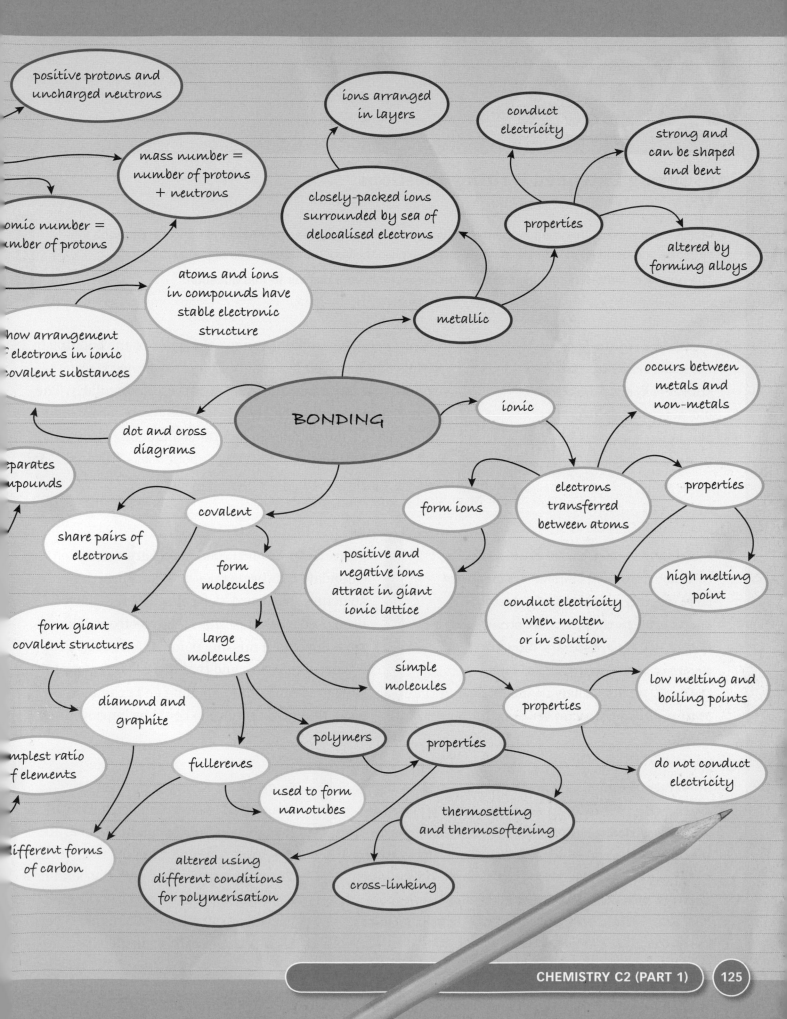

positive protons and uncharged neutrons

mass number = number of protons + neutrons

ions arranged in layers

conduct electricity

strong and can be shaped and bent

omic number = umber of protons

atoms and ions in compounds have stable electronic structure

closely-packed ions surrounded by sea of delocalised electrons

properties

altered by forming alloys

how arrangement f electrons in ionic ovalent substances

metallic

occurs between metals and non-metals

BONDING

ionic

dot and cross diagrams

parates pounds

electrons transferred between atoms

properties

covalent

form ions

share pairs of electrons

form molecules

positive and negative ions attract in giant ionic lattice

conduct electricity when molten or in solution

high melting point

form giant covalent structures

large molecules

simple molecules

low melting and boiling points

mplest ratio f elements

diamond and graphite

polymers

properties

do not conduct electricity

fullerenes

used to form nanotubes

thermosetting and thermosoftening

ifferent forms of carbon

altered using different conditions for polymerisation

cross-linking

properties

# Answering Extended Writing questions

QUESTION

Metals have several important properties that can be explained by their structure and bonding. Describe the structure and bonding in a metal, and use this to explain the important properties of metals.

**The quality of written communication will be assessed in your answer to this question.**

---

**G–E**

Metals are regular and have metallik bonding. This is strong so metals are strong but also bend easily. They can be used in wires and bridges.

Examiner: The candidate knows the name of the bonding in metals (but has spelt it wrongly). The answer needs to explain that the atoms in metals are arranged regularly. The answer demonstrates the knowledge that metals are strong and can bend, but could also mention conducting electricity and heat. The question doesn't ask about uses – the final sentence is irrelevant and scores no marks.

---

**D–C**

Metals are strong, can be bent and conduct electricity . Atoms are arranged close together and regularly. Metallic bonds between the atoms are strong which means metals are strong. There are electrons in the structure so metals conduct. The atoms can slide over each other making the metal bendy.

Examiner: Several good points are made. The candidate lists properties and describes the structure. The word 'delocalised' should have been used when describing the electrons. The most important missing point is that atoms are arranged in layers and that it is the layers that slide over each other. Spelling and grammar are good, though punctuation could be improved.

---

**B–A***

In metals, the atoms are arranged regularly in layers and closely packed. They are surrounded by a sea of delocalised electrons. The force between the atoms and the electrons is called metallic bonding. This is strong so the metal has high tensyl strength The delocalised electrons can move around so it conducts electricity and heat. Metals are flexible and can be bent because the layers slide over each other.

Examiner: A very good, well-structured answer that deals with all the key points. The candidate lists all the properties and explains them well. The only error is that metallic bonding is between electrons and the closely-packed positive ions (not atoms). There is one minor spelling mistake.

# Exam-style questions

**1**

**A01**

**a** Ionic bonding is one of the ways in which atoms join together. Which two of the following statements about ionic bonding are true?

   **i** Electron pairs are shared.

   **ii** Molecules are formed.

   **iii** Positive and negative ions attract each other.

   **iv** It happens between metals and non-metals.

**A01**

**b** Complete this sentence by choosing words from the list below:

Ionic compounds conduct electricity when they are _____ or _____.

solid   liquid   gas   in solution

**2** 99% of carbon atoms are carbon-12, with the symbol $^{12}_{6}C$. However, about 1% of carbon atoms are carbon-13, with the symbol $^{13}_{6}C$. Both have the same atomic number, but different numbers of neutrons.

**A01**

**a** Which word is used to describe these two types of carbon atom?

**A02**

**b** **i** What is the atomic number of both of these carbon atoms?

   **ii** How many neutrons are there in the nucleus of a carbon-13 atom?

**A02**

**c** What can you predict about the relative atomic mass of carbon? Choose the correct answer from the list:

   **i** slightly less than 12

   **ii** exactly 12

   **iii** slightly more than 12

**G–E**

**D–C**

**3** Ammonia, $NH_3$, is manufactured in industry by reacting nitrogen and hydrogen.

**A02**

**a** Balance this equation for the manufacture of ammonia:

$$N_2 + \underline{\ \ }H_2 \rightleftharpoons \underline{\ \ }NH_3$$

**A02**

**b** The predicted yield of ammonia in a reaction was 15 tonnes. However, only 6 tonnes was produced. Calculate the percentage yield.

**A02**

**c** State three reasons why the percentage yield of this reaction is not 100%.

**B–A***

## Extended Writing

**4** Metals are very useful substances.

**A01** What properties do metals have, and what uses do metals have because of them?

**5** GC–MS (gas chromatography–mass

**A01** spectrometry) is an instrumental method used by chemists to analyse substances. Explain why it is used and how it works.

**6** The structures of diamond and

**A02** graphite are shown below:

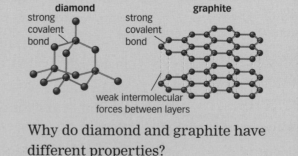

diamond             graphite

strong covalent bond      strong covalent bond

weak intermolecular forces between layers

Why do diamond and graphite have different properties?

**G–E**

**D–C**

**B–A***

# C2 Part 2

# Rates, energy, salts, and electrolysis

## Why study rates, energy, salts, and electrolysis?

Medicines, metals, fireworks, fertilisers – you name it, chemists have helped make it. A huge variety of reaction types make useful products, including acid–base reactions and precipitation reactions. But whatever the reaction type, chemists need answers to two questions: How much energy does the reaction transfer? How fast does it go? They can then work out how to make as much product as possible as quickly as possible.

In this unit, you will find out about rates of reactions, and how to speed them up. You will learn about energy transfers in chemical reactions, too. You will discover how to make salts. Finally, you will take a look at electrolysis.

### You should remember

1. Reactions happen at different speeds. Some, such as rusting, happen slowly. Others, such as explosions, happen very fast indeed.

2. Chemical reactions involve energy transfers, either from or to the surroundings.

3. Energy can be transferred in chemical reactions as heat, light, sound, or electricity.

4. Acids have a pH of less than 7, neutral solutions have a pH of 7, and the pH of an alkaline solution is greater than 7.

5. Ionic substances conduct electricity when melted or in solution.

6. Passing an electric current through an ionic compound that is melted or in solution can break down the ionic compound.

7. Metals above carbon in the reactivity series are extracted from their minerals by electrolysis.

These calcium sulfate crystals are the largest crystals known in the world. They are in the Cave of Crystals in Naica Mine, Chihuahua, Mexico. The crystals formed naturally over millions of years, and were discovered in 2000 after water was pumped out of the mine. In the lab, you can make calcium sulfate crystals from sulfuric acid and calcium oxide.

## Learning objectives

After studying this topic, you should be able to:

✔ explain the meaning of the term rate of reaction

✔ use data, equations, and graphs to calculate reaction rates

▲ Firework reactions happen quickly

A Explain why chemists need to find out about rates of reaction.

B Give two examples of very fast reactions.

## Quick profit

Fernando lives near a lake in Chile. The lake has many salts dissolved in it, including lithium chloride. Fernando wants to set up a factory to make lithium carbonate tablets from the lithium chloride. The tablets treat mood disorders.

The faster Fernando can make the tablets, the sooner he will start making a profit. He experiments to find out how to maximise the **rate of reaction**. The rate of a reaction is a measure of how quickly it happens.

## Fast or slow?

Some reactions, such as a firework exploding, happen in less than a second. Others, such as a metal rusting, may continue for weeks, months, or years.

▲ Rusting reactions happen very slowly

## Following reactions

You cannot find the rate of a reaction from its equation. You need to do experiments to discover how quickly products are made or reactants are used up.

Corinna wants to measure the rate of the reaction of calcium carbonate with hydrochloric acid. The equation for the reaction is:

calcium carbonate + hydrochloric acid → calcium chloride + water + carbon dioxide

$$CaCO_3 + 2HCl → CaCl_2 + H_2O + CO_2$$

Corinna decides to measure the volume of carbon dioxide gas made as the reaction happens. She sets up the apparatus as here on the left.

The reaction starts as soon as both reactants are in the flask. As carbon dioxide gas forms, it pushes out the syringe plunger. Every 30 seconds, for 240 seconds, Corinna reads the total volume of gas that has been produced up until that time.

## Calculating reaction rates

Corinna's first few results are in the table. She plots all her results on a graph, too.

Corinna uses an equation to calculate the rate of reaction over the first 30 seconds:

$$\text{Reaction rate} = \frac{\text{amount of product formed}}{\text{time}} = \frac{15 \text{ cm}^3}{30 \text{ s}} = 0.5 \text{ cm}^3/\text{s}$$

Over the next 30 seconds, from 30 seconds to 60 seconds:

$$\text{Reaction rate} = \frac{(25-15) \text{ cm}^3}{30 \text{ s}} = \frac{10 \text{ cm}^3}{30 \text{ s}} = 0.3 \text{ cm}^3/\text{s}$$

The results of the calculations show that the reaction gets slower as time goes on.

| Time (s) | Volume of gas (cm³) |
|----------|---------------------|
| 0 | 0 |
| 30 | 15 |
| 60 | 25 |
| 90 | 30 |

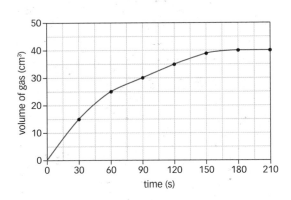

▲ The volume of gas made in the reaction of calcium carbonate and hydrochloric acid. At first the gradient is steep, showing that the rate of reaction is fast. The gradient gets less over time, showing that the reaction slows down.

### Questions

1 Give two examples of very slow reactions.

2 Use results from Corinna's table to calculate the rate of the reaction between 60 seconds and 90 seconds.

3 Draw and label a diagram of the apparatus to measure the rate of reaction for the reaction below. Hydrogen is formed as a gas.

magnesium + hydrochloric acid → magnesium chloride + hydrogen

4 Explain how Corinna's graph shows that the rate of the reaction decreases as time goes by.

5 Use the graph to answer these questions:

(a) How much gas is made in the first 45 seconds?

(b) How long does it take to collect 20 cm³ of gas?

(c) Calculate the rate of reaction between 0 s and 120 s.

**Exam tip**

✔ Practise using the rate of reaction equation. If you are given data about amounts of reactant, you might need to use the equation below, instead of the one given above. Check your units!

$$\text{reaction rate} = \frac{\text{amount of reactant used}}{\text{time}}$$

## Learning objectives

After studying this topic, you should be able to:

✔ recall that chemical reactions happen when particles collide

✔ describe and explain the effect of changing temperature on the rate of reaction

### Fast food

Chris adds chips to boiling water, at 100 °C. They cook in seven minutes. Freya adds chips of the same size to oil, at about 200 °C. They cook more quickly. The chemical reactions that happen in potatoes when they cook are quicker at higher temperatures. The rates of reaction are faster.

◀ Potatoes cook more quickly in oil than in water

▲ Reaction of sodium thiosulfate with HCl

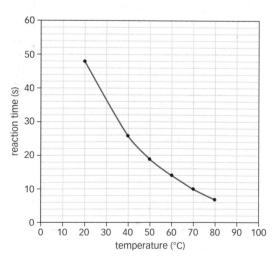

▲ Graph to show how reaction time varies with temperature

### Investigating the effect of temperature

Temperature affects the rates of all reactions. Barney does an investigation to find out more. He uses this equation:

sodium thiosulfate + hydrochloric acid → sodium chloride + water + sulfur dioxide + sulfur

$$Na_2S_2O_3 + 2HCl \rightarrow 2NaCl + H_2O + SO_2 + S$$

Barney's apparatus is shown on the left. The flask on the left shows a mixture of sodium thiosulfate and hydrochloric acid before it has reacted. The flask on the right shows the mixture of products.

Sulfur is insoluble in water, so it makes the reaction mixture turn cloudy. Barney draws a cross on a piece of paper and times how long it takes for the mixture to become so cloudy that he can no longer see the cross. He repeats the experiment at different temperatures.

The graph shows Barney's results. As temperature increases, the reaction happens more and more quickly. The reaction time gets less.

A Name the independent and dependent variables in the investigation.

B Identify the control variables in the investigation.

C Describe what happens to the rate of reaction as temperature increases.

**Key words**

collide, activation energy, successful collision

## When particles collide

Reactions can only happen when reacting particles **collide** with each other. The colliding particles need enough energy, too. The minimum amount of energy that particles need in order to react is the **activation energy**. If two colliding particles have less energy than the activation energy, they will not react.

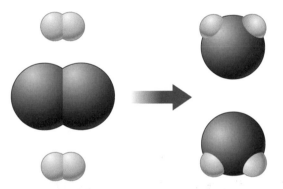

▲ Hydrogen and oxygen molecules can only react to produce water if the molecules collide

## More about particles and rates

The rate of a reaction depends on
- the amount of energy transferred during collisions
- the frequency of collisions.

At low temperatures, particles move relatively slowly. So they do not collide very often. And when they do collide, the amount of energy transferred may not be enough for the particles to react.

At higher temperatures, the particles move faster. They collide more frequently. And, when the particles collide, there is more chance of a **successful collision** (one that leads to a reaction) because faster moving particles transfer more energy when they collide.

**Exam tip**

✔ Remember, faster moving particles collide more frequently and with more energy. So as the reaction rate increases, the reaction time gets less.

### Questions

1 What happens to the rate of a reaction as the temperature decreases?

2 Suggest why the rate of production of water was not measured in the investigation.

↓ E

3 Explain, in terms of particles, what happens to the rate of reaction when the temperature is increased.

↓ C

4 Use the graph to answer these questions:

(a) Explain, in terms of particles, why the rate of reaction doubles between 60 °C and 80 °C.

(b) Estimate the rate of reaction at 30 °C.

↓ A*

# 20: Speeding up reactions – concentration

## Learning objectives

After studying this topic, you should be able to:

✔ describe and explain the effect of changing concentration on rates of reaction

✔ describe and explain the effect of changing pressure on rates of reaction

## Faster and faster

You can speed up reactions by increasing the temperature of the reactants. But this is not the only way to make reactions happen faster. You can also speed up reactions by

- increasing the concentration of reactants in solution
- increasing the pressure of reacting gases
- increasing the surface area of solid reactants
- using a catalyst.

## Rate and concentration

Earl investigates the effect of concentration on reaction rate. He sets up this apparatus.

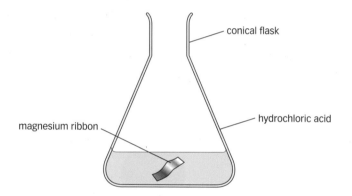

He measures the time for the magnesium to disappear when it reacts with hydrochloric acid of different concentrations. For each test, he uses the same length of magnesium ribbon, and the same volume and temperature of acid.

Earl draws a graph of his results.

**A** Name the independent and dependent variables in Earl's investigation.

**B** List three control variables for the investigation.

**C** What happens to the reaction time as the acid concentration increases?

**D** What happens to the rate of reaction as the acid concentration increases?

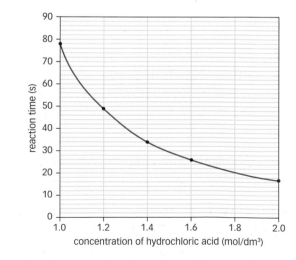

▲ Graph to show how reaction time varies with concentration

# Particles and concentration

The concentration of a solution is a measure of how much solute is dissolved in the solvent. The more concentrated a solution, the greater the number of solute particles that are dissolved in a certain volume of solvent.

 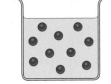

If this represents a 1 mol/dm³ solution of acid ...

then this represents a 2 mol/dm³ solution of the same acid.

▲ There are double the number of acid particles in the same volume of water.

In Earl's investigation, as the acid concentration increased, so the acid particles became more crowded. The frequency of collisions between the acid particles and the magnesium particles increased, so the rate of reaction increased.

# Rate and pressure

Just a tiny spark can set fire to a mixture of hydrogen and oxygen.

$$\text{hydrogen} + \text{oxygen} \rightarrow \text{water}$$
$$2H_2 + O_2 \rightarrow 2H_2O$$

The two gases mix together completely, so their particles collide frequently. If you increase the pressure of the gas mixture, their particles become more crowded. They now collide even more frequently, so the rate of reaction increases.

▲ Hydrogen burns explosively in air

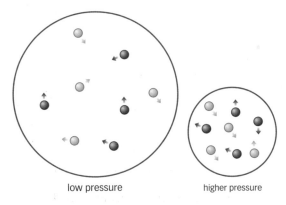

low pressure          higher pressure

▲ Gas particles collide more frequently at higher pressures

## Questions

1 Describe what happens to the rate of reaction involving gases when the pressure is decreased.

2 Describe what happens to the rate of a reaction involving solutions when the concentration is reduced.

3 In Earl's investigation, what was the relationship between reaction time and acid concentration?

4 Explain, in terms of particles, why the rate of reaction increases if the pressure or concentration of the reactants is increased.

**Exam tip**

✔ Increasing the concentration of reactants in solutions, or the pressure of reacting gases, increases the frequency of collisions and so increases the rate of reaction.

## Learning objectives

After studying this topic, you should be able to:

✔ describe and explain the effect of changing surface area on rates of reaction

▲ Powdered milk burns explosively

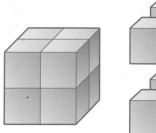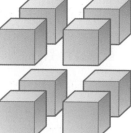

▲ Eight cubes with sides of 1 cm length have the same volume as 1 cube with sides 2 cm length. But the 1 cm cubes have twice the surface area.

## Bang!

TNT (trinitrotoluene) is an explosive. It reacts very fast to release a huge volume of hot gases in a short time. The expanding gases create a shock wave that travels very fast indeed. The shock wave damages objects in its path.

▲ Explosives are used to break the rock face in open cast mines

Powders like flour and custard can cause explosions, too. They burn very easily. This makes them dangerous in factories where they are made or used. The factories have strict safety rules to stop dust escaping into the air inside buildings, and to prevent sparks or naked flames.

## Explaining rate and surface area

A reaction involving a powder happens faster than a reaction involving a lump of the same reactant. A powder has a bigger **surface area** than a lump of the same mass. This is because particles that were inside the lump become exposed on the surface when it is crushed.

Reactant particles must collide for a reaction to happen. The larger the surface area, the greater the frequency of collisions, and so the faster the reaction.

# Investigating rate and surface area

Emily collects data to find out more about how surface area affects reaction rate. She uses this apparatus.

cotton wool

conical flask

calcium carbonate

hydrochloric acid

-0.449

balance

▲ Reaction of calcium carbonate with HCl

The equation for the reaction in the investigation is:

calcium carbonate + hydrochloric acid → calcium chloride + water + carbon dioxide

$$CaCO_3 + 2HCl → CaCl_2 + H_2O + CO_2$$

- First, Emily adds a lump of calcium carbonate to excess hydrochloric acid. Every minute, she measures the loss of mass as carbon dioxide gas escapes from the reacting mixture.
- Next, Emily adds powdered calcium carbonate with the same mass as the lump to excess hydrochloric acid. Again, she measures the loss of mass as carbon dioxide escapes.

Emily plots her results on a graph.

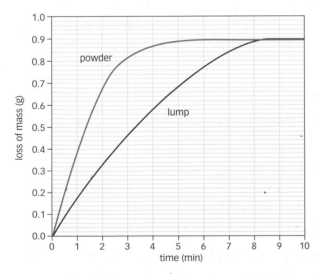

▲ The loss in mass during the reaction of calcium carbonate with hydrochloric acid

## Key words

surface area

**A** Identify the dependent, the independent, and the control variables in the investigation.

**B** Explain how the graph shows that the powder reacts faster than the lumps.

## Exam tip  AQA

✔ Increasing the surface area of solid reactants increases the frequency of collisions and so increases the rate of reaction.

## Questions

1  In general, which react faster, lumps or powders?

2  Explain why increasing the surface area of a solid reactant increases the rate of a reaction.

3  Use the graph to answer these questions:
   (a) When did each of the reactions finish?
   (b) Which reaction was faster during the first two minutes?

4  Calculate the mean (average) rate of reaction for each reaction shown in the graph.

5  Explain why powders are dangerous in factories.

## Learning objectives

After studying this topic, you should be able to:

- ✔ explain what catalysts are and what they do
- ✔ explain why catalysts are important in industry

## Cats in cars

There are millions of cars in the UK. And each one is a source of pollution. To tackle this problem, car exhaust systems are fitted with catalytic converters, or 'cats'. In a catalytic converter, chemical reactions convert harmful exhaust gases into less harmful ones. For example:

$$\text{carbon monoxide} + \text{oxygen} \rightarrow \text{carbon dioxide}$$
$$2CO + O_2 \rightarrow 2CO_2$$

This reaction can happen on its own, but under the conditions in an exhaust system the reaction is too slow to get rid of carbon monoxide before it goes into the air. The catalytic converter speeds up the reaction. But how? What's inside a cat?

## Catalysts

You can speed up reactions by increasing

- the temperature
- the concentration, if solutions are involved
- the pressure, if gases are involved
- the surface area, if solids are involved.

You can also use a **catalyst** to make a reaction faster. A catalyst is a substance that increases the rate of a chemical reaction without being used up in the reaction. So if you add 1 g of a catalyst to a reaction mixture, there is still 1 g of it left when the reaction has finished.

Catalytic converters contain two or three precious metals as catalysts, usually chosen from platinum, rhodium, and palladium.

## Just a little

In July 2010, just one gram of platinum cost about £32. Luckily, a small amount of catalyst will **catalyse** the reaction between large amounts of reactants, so a typical catalytic converter needs only 3 g of platinum catalyst.

Catalysts are specific to particular reactions. A substance that acts as a catalyst for one reaction may not work as a catalyst for another reaction.

▲ Catalytic converters in cars convert harmful exhaust gases into less harmful ones

**A** What is a catalyst?

**B** Describe one way in which platinum is used as a catalyst.

## Catalysts in the lab

If you've ever bleached your hair, you've probably used hydrogen peroxide solution.

Hydrogen peroxide is usually diluted in water, where it breaks down *very* slowly:

$$\text{hydrogen peroxide} \rightarrow \text{water} + \text{oxygen}$$
$$2H_2O_2 \rightarrow 2H_2O + O_2$$

Powdered manganese(IV) oxide, $MnO_2$, catalyses this decomposition reaction. If you add a small amount of manganese(IV) oxide to hydrogen peroxide solution, lots of oxygen quickly forms.

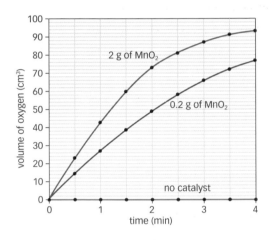

Adding a lot more catalyst makes little difference to the rate at which hydrogen peroxide breaks down

▲ Hydrogen peroxide decomposes when a manganese(IV) oxide catalyst is added

## Questions

1 Give an example of a catalyst and the reaction it catalyses.

2 Explain why catalysts are important in industry.

3 Use the graph to answer these questions:

   (a) What was the effect on the rate of reaction of adding manganese(IV) oxide?

   (b) Explain which reaction was faster in the first two minutes.

   (c) How can you tell that the reactions were not complete after four minutes?

4 Calculate the mean rate of reaction for each reaction shown in the graph.

5 Suggest why adding ten times as much manganese(IV) oxide did not increase the rate of reaction by ten times.

### Key words

catalyst, catalyse

### Exam tip   AQA

✔ Catalysts are useful in the chemical industry. Without them, many industrial processes would be too slow to be profitable. For example, iron is used in the Haber process, which makes ammonia for fertilisers and explosives. The iron catalyst increases the rate of the reaction between nitrogen and hydrogen. So more ammonia is made in a shorter time.

## Learning objectives

After studying this topic, you should be able to:

✔ explain what exothermic reactions are, and give examples of them

## Key words

**exothermic reaction**

## Firework fun

Fabulous fireworks light up the sky at Diwali and the new year. Chemical reactions in fireworks transfer energy to the surroundings as heat, light, and sound. Reactions that transfer energy to the surroundings are called **exothermic reactions**.

▲ Firework display

**A** What is an exothermic reaction?

**B** State three forms of energy that can be transferred by chemical reactions.

## Combustion reactions

Exothermic reactions do not happen only in fireworks. All combustion reactions transfer energy to the surroundings, mainly as heat and light. The heat they transfer is useful for cooking, heating homes, and generating electricity. Power stations burn coal to heat water into steam. The steam turns turbines which generate electricity.

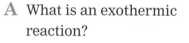

## More exothermic reactions

Neutralisation reactions are exothermic, too. Daisha adds 50 cm³ of dilute hydrochloric acid to 50 cm³ of sodium hydroxide solution. She records the temperatures before and after the reaction.

| Substance | Temperature (°C) |
|---|---|
| Hydrochloric acid, before mixing | 20 |
| Sodium hydroxide solution, before mixing | 20 |
| Reaction mixture, immediately after mixing | 49 |
| Reaction mixture, 1 hour after mixing | 20 |

Hydrochloric acid and sodium hydroxide react in a neutralisation reaction:

$$\text{hydrochloric acid} + \text{sodium hydroxide} \rightarrow \text{sodium chloride} + \text{water}$$

$$\text{HCl} + \text{NaOH} \rightarrow \text{NaCl} + \text{H}_2\text{O}$$

The reaction gives out energy. At first, this energy heats up the reacting mixture. Then heat is transferred from the mixture to the surroundings, and the mixture cools to room temperature.

Some other types of reaction are exothermic, including:
- many oxidation reactions – for example the reaction in which potassium chlorate oxidises the glucose sugar in a jelly baby, causing it to burst into flames and make a screaming sound

◀ Jelly babies

- displacement reactions – for example when zinc reacts with copper sulfate solution to make copper and zinc sulfate.

## Using exothermic reactions

Hand warmers use exothermic changes to produce heat. In one type, iron is quickly oxidised when you activate the hand warmer. The reaction is exothermic, so it transfers heat to your hands.

Self-heating coffee cans have two compartments. One contains cold coffee. The other contains reactants which react together in an exothermic reaction. This transfers heat to the coffee. Delicious!

▲ Glow worm at night

### Did you know...?

Chemical reactions in glow worms transfer light energy to the surroundings.

### Questions

1 Name three types of reaction which are exothermic.

2 Give two examples of exothermic reactions which are useful.

3 Describe what happens to the temperature of a reaction mixture during and after an exothermic reaction.

4 Suggest how you could find out whether the reaction of magnesium with dilute hydrochloric acid is exothermic or not. What results would you expect if the reaction is exothermic?

5 Explain how Daisha's data show that the neutralisation reaction is exothermic.

↓ E

↓ C

↓ A*

## Learning objectives

After studying this topic, you should be able to:

✔ recall that, in chemical reactions, energy can be transferred from or to the surroundings

✔ explain what endothermic reactions are, and give examples of them

✔ recall that if a reaction is exothermic in one direction it is endothermic in the opposite direction

---

**A** What is an endothermic reaction?

**B** Give two examples of endothermic reactions.

---

## Sherbet fizz

What happens when you eat a sherbet? It fizzes in your mouth, and your tongue feels cold. The fizzing happens when two sherbet ingredients – sodium hydrogencarbonate and citric acid – react together. This is an **endothermic reaction**. It takes in energy from the surroundings, in this case in the form of heat from your tongue.

▲ Sherbet sweets

The sherbet reaction also happens if you put the sherbet in water. You can tell it is endothermic because the temperature decreases.

## More endothermic reactions

Here are some more examples of endothermic processes:

▲ Energy must be supplied to get thermal decomposition reactions to happen. Here, copper carbonate is decomposing to make copper oxide and carbon dioxide.

▲ When ammonium nitrate dissolves in water, the temperature falls so much that a drop of water can freeze a container to a block of wood

## Using endothermic reactions

Sports injury packs relieve pain on the football field and hockey pitch. Some are based on an endothermic reaction – activating them starts off a chemical reaction that takes in heat energy from the injured leg or arm.

## Reversible reactions

All chemical reactions involve energy transfers. If a reversible reaction is exothermic in one direction, it is endothermic in the opposite direction. The same amount of energy is transferred in each direction:

- Pawel takes some blue copper sulfate crystals. The crystals are **hydrated** – they contain water. The formula of hydrated copper sulfate is $CuSO_4.5H_2O$. Pawel heats the crystals. A white powder forms. The white powder is **anhydrous** copper sulfate, $CuSO_4$. It has no water in it. This process is endothermic – it takes in heat energy from the surroundings.
- Pawel waits for the white powder to cool. Then he adds a few drops of water to it. Blue crystals form again, and heat energy is given out. The process is exothermic.

You can summarise Pawel's reactions like this:

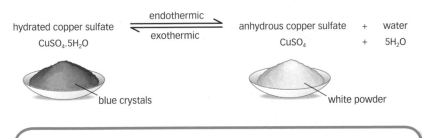

hydrated copper sulfate $\underset{\text{exothermic}}{\overset{\text{endothermic}}{\rightleftarrows}}$ anhydrous copper sulfate + water

$CuSO_4.5H_2O$ ⇌ $CuSO_4$ + $5H_2O$

blue crystals · white powder

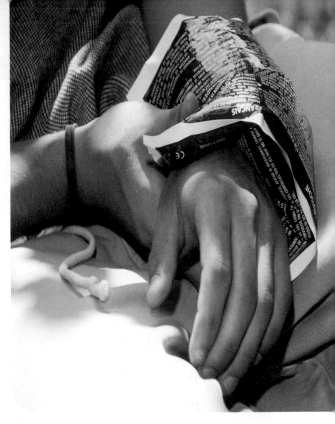

▲ Sports injury packs cool damaged muscles

## Questions

1 Give an example of a useful endothermic reaction.

2 Draw up a table to summarise the differences between endothermic and exothermic reactions. Include ideas about temperature changes and energy transfer.

3 Draw diagrams to summarise what Pawel did in his experiment, and to show the direction in which the reaction is exothermic, and the direction in which it is endothermic.

**Exam tip** **AQA**

✔ Think of a fire exit sign to help you remember which way round exothermic and endothermic reactions go. A fire is hot and you go out of an exit. Exothermic reactions get hot as they give energy out.

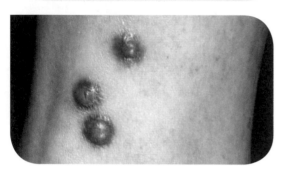

▲ Ant stings are acidic, like bee stings

### Did you know...?

Ant stings are acidic, like bee stings, but wasp stings are alkaline.

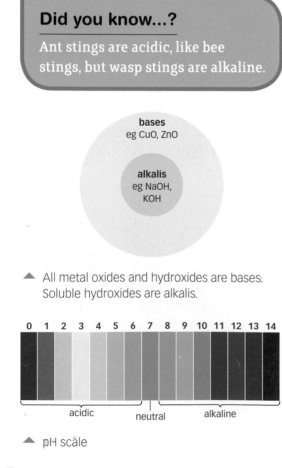

**bases**
eg CuO, ZnO

**alkalis**
eg NaOH, KOH

▲ All metal oxides and hydroxides are bases. Soluble hydroxides are alkalis.

| 0 | 1 | 2 | 3 | 4 | 5 | 6 | 7 | 8 | 9 | 10 | 11 | 12 | 13 | 14 |

acidic    neutral    alkaline

▲ pH scale

## Bee sting

Harry gets stung by a bee. It hurts! He puts toothpaste on the sting. Soon, the sting is less painful. The toothpaste has neutralised – 'cancelled out' – the acid in the bee sting.

▲ Bee stings are acidic

## Alkalis and bases

Toothpaste is not the only substance that neutralises acids. All bases and alkalis do:

- **Bases** are metal oxides and hydroxides, for example, copper oxide.
- Hydroxides that are soluble are called **alkalis**. Sodium hydroxide and potassium hydroxide are alkalis.

Charged particles called **hydroxide ions**, OH⁻(aq), make solutions alkaline. The (aq) shows that the hydroxide ion is dissolved in water.

## The pH scale

The **pH scale** measures the acidity or alkalinity of a solution:

- A solution of pH 7 is neutral.
- A solution with a pH of less than 7 is acidic.
- A solution with a pH of more than 7 is alkaline.

You can use indicators such as Universal indicator to show the pH of a solution.

A Give the formula of the particle that makes solutions alkaline.

B A solution has a pH of 6.8. Is the solution acidic, alkaline, or neutral?

## Neutralisation

Hydroxide ions make solutions alkaline. But what makes solutions acidic? The answer is **hydrogen ions**, $H^+(aq)$. Hydrogen ions only make solutions acidic if they are dissolved in water.

When sodium hydroxide neutralises hydrochloric acid, the products are sodium chloride and water:

$$\text{hydrochloric acid} + \text{sodium hydroxide} \rightarrow \text{sodium chloride} + \text{water}$$

$$HCl(aq) + NaOH(aq) \rightarrow NaCl(aq) + H_2O(l)$$

In this reaction, hydrogen ions from the hydrochloric acid react with hydroxide ions from the sodium hydroxide to produce water. You can represent the reaction – and all **neutralisation reactions** – like this:

$$H^+(aq) + OH^-(aq) \rightarrow H_2O(l)$$

The (l) shows that the water is liquid. Sodium ions and chloride ions play little part in the reaction of hydrochloric acid with sodium hydroxide.

## Amazing ammonia

Like metal hydroxides, ammonia dissolves in water to produce an alkaline solution. The formula of ammonia is $NH_3$. In water, it forms hydroxide ions like this:

$$\text{ammonia(g)} + \text{water(l)} \rightarrow \text{ammonium hydroxide(aq)}$$

$$NH_3(g) + H_2O(l) \rightarrow NH_4OH(aq)$$

When dissolved in water, the ammonium and hydroxide ions are separated and surrounded by water molecules. So ammonium hydroxide solution contains $NH_4^+(aq)$ ions and $OH^-(aq)$ ions. The $OH^-(aq)$ ions make the solution alkaline.

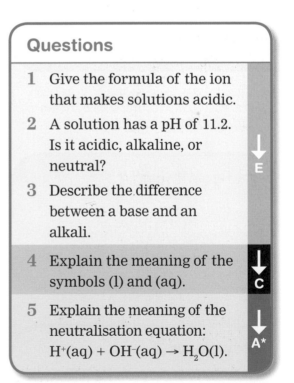

**Questions**

1 Give the formula of the ion that makes solutions acidic.

2 A solution has a pH of 11.2. Is it acidic, alkaline, or neutral?

3 Describe the difference between a base and an alkali.

4 Explain the meaning of the symbols (l) and (aq).

5 Explain the meaning of the neutralisation equation: $H^+(aq) + OH^-(aq) \rightarrow H_2O(l)$.

▲ *Seizure*, 2008. To create this art installation, the artist allowed copper sulfate crystals to grow in a London flat.

### Did you know...?

You can use copper sulfate to treat aquarium fish for parasitic infections. But don't add too much – copper ions are toxic to fish!

### Key words

salt, soluble, crystallisation

### Seizure

In 2008, artist Roger Hiorns poured 75 000 litres of copper sulfate solution through a hole in the ceiling of an abandoned London flat. Crystals began to grow. A few weeks later, Roger pumped out the remaining solution. The inside of the flat was covered in stunning jewel-like crystals, glinting and other-worldly. Roger named his piece of art 'Seizure'.

### Salts

Copper sulfate is an example of a **salt**. Of course it is different from the substance we normally call salt – that's sodium chloride. And you certainly can't eat copper sulfate – swallowing just 1 g would make you vomit and turn yellow.

A salt is a compound that contains metal ions and that can be made from an acid.

### Which acid?

Different acids make different types of salt:
- Hydrochloric acid (HCl) makes chlorides, for example sodium chloride.
- Sulfuric acid ($H_2SO_4$) makes sulfates, for example copper sulfate.
- Nitric acid ($HNO_3$) makes nitrates, for example zinc nitrate.

### Metal ions for salts

Several types of substances can supply metal ions to salts, including:
- metals, such as magnesium
- insoluble bases, such as copper oxide
- alkalis, such as sodium hydroxide.

> **A** What is a salt?
> **B** Suggest which metal and acid you could use to make magnesium chloride.

# Making a soluble salt from an insoluble base

Copper sulfate dissolves in water, so chemists call it a **soluble** salt. You can make copper sulfate by reacting sulfuric acid with copper oxide, an insoluble base:

copper oxide + sulfuric acid → copper sulfate + water

$$CuO(s) + H_2SO_4(aq) \rightarrow CuSO_4(aq) + H_2O(l)$$

The (s) in the symbol equation means that the copper oxide is solid.

Here's how to make the copper sulfate:

**Step 1** Add a little copper oxide to a sample of dilute sulfuric acid. Keep adding copper oxide until no more reacts. There will now be a blue solution of copper sulfate mixed with some black copper oxide powder that has not reacted.

**Step 2** Filter to remove the unreacted copper oxide powder.

**Step 3** Heat the blue solution over a water bath, until about half its water has evaporated.

**Step 4** Leave the solution to stand for a few days. Crystals will slowly form. This is **crystallisation**.

**Exam tip** AQA

✔ You need to be able to suggest methods to make a named soluble salt.

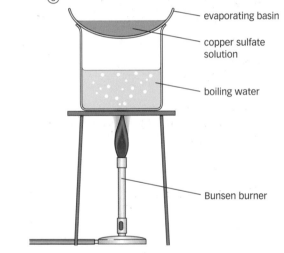

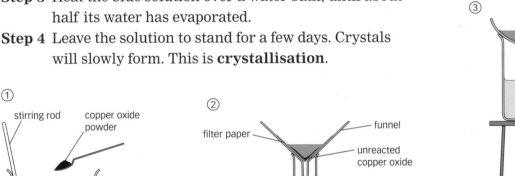

① stirring rod — copper oxide powder — dilute sulfuric acid

② filter paper — funnel — unreacted copper oxide — copper sulfate solution

③ evaporating basin — copper sulfate solution — boiling water — Bunsen burner

## Questions

1 Name the type of salt formed from nitric acid.

2 Explain the meaning of the word crystallisation.

3 Suggest which insoluble base (metal oxide) and which acid you could use to make the salts listed below. Write word equations for the reactions:

   (a) zinc nitrate       (b) copper chloride       (c) magnesium sulfate.

4 Describe how to make zinc chloride from zinc oxide and hydrochloric acid. Include a balanced equation for the reaction. Useful formulae: $ZnO$, $ZnCl_2$.

↓ E
↓ C
↓ A*

## Magnesium sulfate

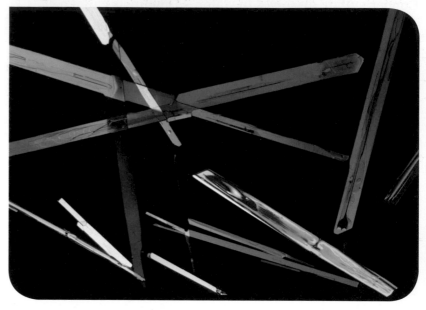

▲ Magnesium sulfate crystals

Magnesium sulfate forms beautiful crystals. It is useful, too. It helps potato, rose, and tomato plants grow well. It is also an effective laxative.

Magnesium sulfate occurs naturally in the Earth's crust. You can also make the salt in the laboratory.

### Making a soluble salt from a metal

Magnesium sulfate forms in the reaction of magnesium with sulfuric acid:

magnesium + sulfuric acid → magnesium sulfate + hydrogen

$$Mg(s) + H_2SO_4(aq) \rightarrow MgSO_4(aq) + H_2(g)$$

The (g) shows that hydrogen is a gas.

> **A** Give the meanings of the symbols (s), (l), (g), and (aq).

Here's how to make the magnesium sulfate:
- Add a small piece of magnesium ribbon to a sample of dilute sulfuric acid. Keep on adding magnesium until the bubbling stops and there is a little solid magnesium in the colourless solution of magnesium sulfate.
- Filter to remove unwanted magnesium ribbon.
- Heat the colourless solution over a water bath, until about half its water has evaporated.
- Allow the solution to crystallise by leaving it to stand for a few days.

Not all metals react with acids to form salts. Some, like copper, are not reactive enough. Other metals, for example sodium, are too reactive – it is dangerous to add sodium metal to dilute acid in the laboratory.

> **B** Suggest which metal and acid you could use to make zinc nitrate.

## Making a soluble salt from an alkali

Sodium chloride can be made in the lab by burning sodium in chlorine gas. But it is safer and simpler to make this salt by reacting dilute hydrochloric acid with sodium hydroxide solution:

$$\text{sodium hydroxide} + \text{hydrochloric acid} \rightarrow \text{sodium chloride} + \text{water}$$

$$\text{NaOH(aq)} + \text{HCl(aq)} \rightarrow \text{NaCl(aq)} + \text{H}_2\text{O(l)}$$

The pictures below show how to make the sodium chloride.

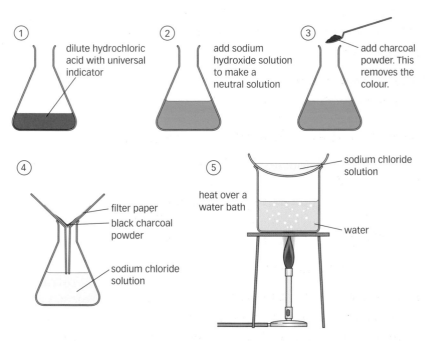

① dilute hydrochloric acid with universal indicator

② add sodium hydroxide solution to make a neutral solution

③ add charcoal powder. This removes the colour.

④ filter paper
black charcoal powder
sodium chloride solution

⑤ heat over a water bath
sodium chloride solution
water

▲ Instructions for making sodium chloride

### Questions

1 Name the salt made when you react magnesium with sulfuric acid.

2 When making magnesium chloride from a metal and an acid, how can you tell when you have added enough magnesium to the hydrochloric acid?

3 When making potassium chloride from an acid and alkali, how can you tell when you have added enough potassium hydroxide to the hydrochloric acid?

4 Suggest which alkali and which acid you could use to make the salts listed below. Write word equations for the reactions:

   (a) potassium nitrate

   (b) sodium nitrate.

5 Describe how to make magnesium chloride ($\text{MgCl}_2$) from a metal and an acid. Include a balanced equation for the reaction.

# : Precipitation reactions

## Learning objectives

After studying this topic, you should be able to:

- ✔ suggest the substances needed to make an insoluble salt
- ✔ give examples of how precipitation reactions are useful

## Key words

**precipitate, precipitation reaction, ionic equation**

## Exam tip

- ✔ You need to be able to suggest pairs of solutions that react to make a named insoluble salt in a precipitation reaction.

## Chemical magic?

Lead nitrate solution is colourless and transparent. So is potassium iodide solution. But if you mix the solutions together you make something quite different – a bright yellow **precipitate**. A precipitate is a suspension of small solid particles, spread throughout a liquid or solution. It makes the mixture look cloudy.

◀ Lead iodide forms as a precipate if you mix lead nitrate and potassium iodide solutions

If you filter the yellow mixture, a yellow solid collects on the filter paper. A colourless solution drips into the flask.

## Explaining precipitation

The equation for the reaction that makes the yellow precipitate is:

lead nitrate $+$ potassium iodide $\rightarrow$ lead iodide $+$ potassium nitrate

$$Pb(NO_3)_2(aq) + 2KI(aq) \rightarrow PbI_2(s) + 2KNO_3(aq)$$

The symbol equation shows that the lead iodide forms as a solid. It is insoluble in water. So the yellow precipitate is lead iodide. The reaction is a **precipitation reaction**, since it forms a precipitate.

The potassium and nitrate ions do not change in the reaction. You can use an **ionic equation** to summarise the reaction. It shows only the ions that take part.

$$Pb^{2+}(aq) + 2I^-(aq) \rightarrow PbI_2(s)$$

**A** What is a precipitation reaction?

## Making other insoluble salts

You can make other insoluble salts in precipitation reactions, too. For example:

| barium chloride | + | sodium sulfate | → | barium sulfate | + | sodium chloride |

$$BaCl_2(aq) + Na_2SO_4(aq) \rightarrow BaSO_4(s) + 2NaCl(aq)$$

## Using precipitation reactions

Precipitation reactions are useful for removing unwanted ions from solution. For example, waste water may contain poisonous lead, manganese, or chromium ions. These can be removed by adding a solution that contains negative ions that form a precipitate with the metal ions. For example:

$$Cr^{3+}(aq) + 3OH^-(aq) \rightarrow Cr(OH)_3(s)$$

The hydroxide ions are supplied by calcium hydroxide solution. Filtering removes the chromium(III) hydroxide precipitate.

## Working out formulae

If you know the charges of the ions in a compound, you can work out its formula. Overall, ionic compounds have no electrical charge. The positive and negative charges cancel each other out. So formulae must show equal numbers of positive and negative charges. For example, to find the formula of barium chloride:

- The charges on the ions are $Ba^{2+}$ and $Cl^-$.
- A neutral compound needs two $Cl^-$ ions for every one $Ba^{2+}$ ion.
- The formula of the compound is $BaCl_2$.

The table shows the charges on some ions.

| Charge | Examples of ions with this charge | | | |
|--------|------|------|------|------|
| –2 | $O^{2-}$ | $S^{2-}$ | $SO_4^{2-}$ | |
| –1 | $Cl^-$ | $Br^-$ | $I^-$ | $NO_3^-$ |
| +1 | $Li^+$ | $Na^+$ | $K^+$ | |
| +2 | $Mg^{2+}$ | $Ca^{2+}$ | $Ba^{2+}$ | |
| +3 | $Al^{3+}$ | | | |

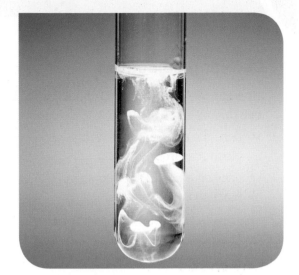

▲ Barium sulfate precipitate

**B** Name two solutions that react to make a precipitate of barium sulfate.

## Questions

1 What is a precipitate?

2 Write a word equation for the formation of lead iodide from potassium iodide and lead nitrate solutions.  ↓ E

3 Write down the formulae of magnesium chloride, potassium sulfate, and aluminium iodide.  ↓ C

4 Suggest two solutions you could mix to make a precipitate of copper hydroxide.

5 Suggest a solution you could mix with silver nitrate solution to make a precipitate of silver chloride.  ↓ A*

## Learning objectives

After studying this topic, you should be able to:

✔ describe what happens at the electrodes in electrolysis
✔ use half equations to represent reactions at electrodes

## Key words

**electrolysis**, **electrolyte**, **reduction**, **oxidation**, **half equation**

## Extraordinary element

What do the objects in the pictures have in common?

All use the element lithium, or its compounds. Many electronic goods are powered by lithium batteries. Lithium stearate grease is a useful vehicle lubricant. Submarines use lithium peroxide to remove carbon dioxide from the air.

You can't just dig lithium out of the ground. It is much too reactive to exist on its own. Most lithium is produced by the **electrolysis** of lithium salts.

## What is electrolysis?

When an ionic compound is melted, or dissolved in water, its ions are free to move about within the liquid or solution. Passing an electric current through the liquid or solution breaks down the ionic compound into simpler substances. This is electrolysis. The solution of the substance that is broken down is called the **electrolyte**.

During the electrolysis of molten (melted) lead bromide:
- Positive lead ions move towards the negative electrode.
- Negative bromide ions move towards the positive electrode.

negative electrode ⊖   ⊕ positive electrode

Key:
● lead ion, $Pb^{2+}$
● bromide ion, $Br^-$

Note: This is a simplified diagram. In fact, the whole liquid is made up of lead ions and bromide ions only.

▲ Electrolysis of molten lead bromide

**A** What is electrolysis?

**B** In electrolysis, what type of ion moves towards the negative electrode?

# What happens at the electrodes?

- At the negative electrode, positively charged ions gain electrons. This is **reduction**. In the example, lead ions gain electrons. They are reduced.
- At the positive electrode, negatively charged ions lose electrons. This is **oxidation**. In the example, bromide ions lose electrons. They are oxidised.

Oxidation is not just about adding oxygen – a better definition is that oxidation is the loss of electrons.

Chemists use **half equations** to show what happens at the electrodes during electrolysis. Electrons are represented by $e^-$. You can balance a half equation by adding or subtracting electrons until the total charge on each side is equal.

For example, in the electrolysis of lead bromide:
- At the negative electrode, $Pb^{2+} + 2e^- \rightarrow Pb$
- At the positive electrode, $2Br^- - 2e^- \rightarrow Br_2$
  This is also written as $2Br^- \rightarrow Br_2 + 2e^-$.

The positive electrode half equation also shows that, at this electrode, bromide ions lose electrons. The resulting atoms join together in pairs to form bromine molecules, $Br_2$.

## Questions

1 In the electrolysis of lead bromide, which ion moves to the negative electrode?

2 Explain why electrolysis only breaks down melted or dissolved ionic compounds.

3 Give two examples of uses of electrolysis.

4 Write half equations to show what happens at the electrodes during the electrolysis of copper chloride solution. Ion formulae: $Cu^{2+}$ and $Cl^-$

5 Suggest economic and environmental reasons for using renewably generated electricity for the electrolysis of water to make hydrogen fuel.

**Exam tip**

- ✔ Negative ions lose electrons at the positive electrode. This is oxidation.
- ✔ Positive ions gain electrons at the negative electrode. This is reduction.
- ✔ Use OIL RIG to help you remember – Oxidation Is Loss, Reduction Is Gain.

### Key words

electroplating

## All that glistens is not gold

Raj had his personal music player covered in gold. It looks fantastic.

▲ This personal music player is covered in gold

Linda has a shiny motorbike. It is made mainly from steel. But steel rusts easily. So during its manufacture, the bike was coated with a layer of chromium metal. Chromium resists corrosion, so it protects the bike.

▲ This motorbike has a thin layer of chromium on its surface

Food cans are made from steel. They have a very thin coating of tin. This stops the steel rusting. You couldn't make cans from tin alone – they would be too expensive.

How were these objects coated with gold, chromium, or tin? Read on to find out.

### Did you know...?

You can't electroplate steel directly with chromium, because chromium doesn't stick well to steel. You have to electroplate steel first with copper, then with nickel, and finally with chromium.

# Electroplating

The music player, bike, and cans were all coated by a process called **electroplating**. Electroplating happens in electrolysis cells, like the one shown on the right.

> **A** What is electroplating?
>
> **B** Give two reasons for electroplating objects.

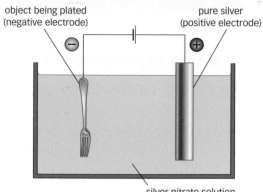

How to electroplate a fork with silver

## What happens at the electrodes?

In the electrolysis cell:

- Positive silver ions move to the negative electrode. Here, they gain electrons to form silver metal. The ions have been reduced.
- At the positive electrode, silver ions from the electrode lose electrons and go into solution. The silver has been oxidised.

You can write half equations for the reactions at the electrodes:

- Negative electrode $Ag^+ + e^- \rightarrow Ag$
- Positive electrode $Ag - e^- \rightarrow Ag^+$

## Questions

1 Suggest why Raj got his music player electroplated with gold.

2 Suggest a reason for electroplating steel cutlery with silver.

3 Suggest a reason for electroplating a car bumper with chromium.

4 Explain why, in electroplating, the object to be coated must always be the negative electrode of an electrolysis cell.

5 Draw a diagram to show how you could set up a nickel electrode and a nickel sulfate solution to electroplate a paper clip with nickel.

**Exam tip**

- ✔ The object to be electroplated must always be the negative electrode of an electrolysis cell.

# 31: Using electrolysis – 1

▲ The properties of aluminium make it suitable for take-away food containers

## Awesome aluminium

There is huge demand for aluminium – more than 50 million tonnes of the metal was produced in 2006. Aluminium conducts electricity well, resists corrosion, and has a low density. These properties make it perfect for planes, packaging, and power lines. In 2010, a tonne of aluminium cost about £670, compared to about £280 for a tonne of steel.

▲ Aluminium ingots are valuable

Aluminium compounds are everywhere. Aluminium is the most abundant metal in the Earth's crust. So why is aluminium so expensive?

## Extracting aluminium

Aluminium exists naturally mainly as aluminium oxide, in **bauxite** ore. Aluminium is relatively high in the reactivity series. Its oxide cannot be reduced by heating with carbon. So aluminium is extracted by electrolysis. This is what makes aluminium expensive – an aluminium plant producing about 120 000 tonnes a year needs more than 200 megawatts of electricity, enough to meet the needs of a small town.

Here's how to extract aluminium from bauxite:

- Remove impurities from the ore to get pure aluminium oxide.
- Dissolve the pure aluminium oxide in molten **cryolite**. The mixture melts at around 950 °C. Pure aluminium oxide melts at 2070 °C. This temperature is too high for it to be used alone.
- Pour the liquid aluminium oxide and cryolite mixture into a huge electrolysis cell, as shown in the diagram below.
- Pass a current of 100 000 A through the liquid mixture.
    - Aluminium ions move to the negative electrode. Here, they gain electrons and form liquid aluminium metal.
    - Oxide ions move to the positive electrode. Here, they give up electrons to form oxygen gas. The oxygen reacts with the carbon electrode, making carbon dioxide gas.
- The half equation for the reaction at the negative electrode is $Al^{3+} + 3e^- \rightarrow Al$
- The half equation for the reaction at the positive electrode is $2O^{2-} - 4e^- \rightarrow O_2$

**Key words**

bauxite, cryolite

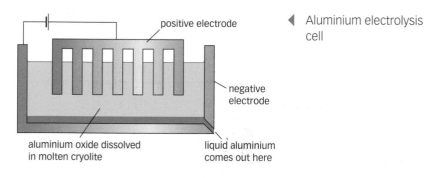

positive electrode

negative electrode

aluminium oxide dissolved in molten cryolite

liquid aluminium comes out here

◄ Aluminium electrolysis cell

**Exam tip**    AQA

✔ Remember – the purpose of the cryolite is to reduce the temperature of the electrolysis cell. Aluminium forms at the negative electrode. Carbon dioxide forms at the positive electrode.

## Questions

1. Name the substances formed at the positive and negative electrodes in the production of aluminium from aluminium oxide.

2. Name the ore from which most aluminium is extracted.

3. Describe what happens at the positive electrode during the electrolysis of aluminium oxide.

↓ E

↓ C

4. Write half equations for the reactions that occur at the positive and negative electrodes during the production of aluminium.

5. Suggest social, economic, and environmental benefits of recycling aluminium.

↓ A*

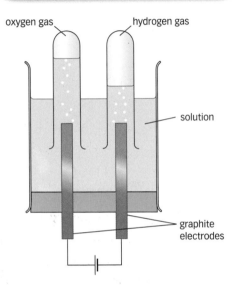

oxygen gas    hydrogen gas

solution

graphite electrodes

▲ Passing electricity through calcium sulfate solution produces oxygen and hydrogen gases

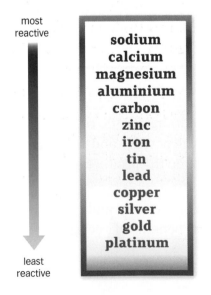

most reactive

sodium
calcium
magnesium
aluminium
carbon
zinc
iron
tin
lead
copper
silver
gold
platinum

least reactive

▲ Part of the reactivity series. The reactivity series lists metals in order of how vigorously they react with substances such as oxygen and water.

## Products of electrolysis

If you electrolyse molten lead bromide, there is only one possible product at each electrode:

• lead at the negative electrode

• bromine at the positive electrode.

But for ionic compounds that are dissolved in water, predicting the electrolysis products is not quite so simple. Water takes part in electrolysis reactions, too.

## Predicting products

Isabelle passed an electric current through some solutions, and identified the products at the electrodes.

The table summarises some of her results.

| Solution | Product at negative electrode | Product at positive electrode |
|---|---|---|
| potassium iodide | hydrogen | iodine |
| magnesium bromide | hydrogen | bromine |
| copper nitrate | copper | oxygen |
| sodium carbonate | hydrogen | oxygen |
| calcium sulfate | hydrogen | oxygen |

There is a pattern in the results.

• At the negative electrode:
  – The metal is produced if it is low in the reactivity series, like copper.
  – Hydrogen gas forms if the metal is above copper in the reactivity series. The hydrogen comes from the water.
• At the positive electrode:
  – If halide ions are present in solution, halogens will be produced.
  – If carbonate, sulfate, or nitrate ions are in the solution, oxygen is produced. The oxygen comes from the water.

> **A** Predict the products of the electrolysis of sodium chloride solution.
>
> **B** Predict the products of electrolysis of copper sulfate solution.

# The electrolysis of sodium chloride solution

A whole industry has built up around the electrolysis of concentrated sodium chloride solution, or **brine**.

Passing electricity through brine produces these products:
- hydrogen gas at the negative electrode
- chlorine gas at the positive electrode.

A solution of sodium hydroxide also forms.

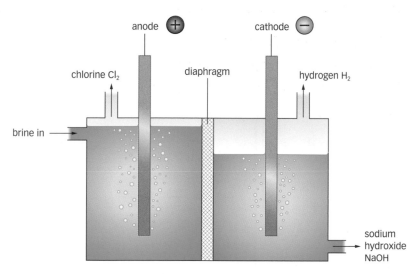

▲ The electrolysis of brine. The diaphragm keeps the chlorine away from the sodium hydroxide solution.

## Using products from brine

The products of electrolysis of brine have many uses.
For example:
- Sodium hydroxide is used to make soap.
- Chlorine is used to make bleach and plastics, and to kill bacteria in drinking water and swimming pools.
- Hydrogen is used to make margarine and ammonia.

▲ Cells for the electrolysis of brine

**Key words**

brine

**Questions**

1 Name the three products of the electrolysis of brine.

2 Give one use for each of the products of the electrolysis of brine.

↓ E

3 Predict the products of the electrolysis of potassium bromide solution.

4 Predict the products of the electrolysis of melted potassium bromide.

↓ C

5 Write a half equation for the reaction that happens at the positive electrode during the electrolysis of brine.

6 Write half equations for the reactions that happen at the electrodes during the electrolysis of copper bromide solution.

↓ A*

# Course catch-up

## Revision checklist

- Rate of reaction is the change in amount of product (or reactant) per second.
- Reactions are faster at higher temperatures, at high concentrations and high pressure, and with powdered solids.
- Activation energy is the minimum amount of energy particles need to react.
- Catalysts increase the rate of a reaction but are unchanged at the end. They reduce the cost of industrial reactions.
- Some reactions (eg combustion) are exothermic (transfer energy to the surroundings).
- Some reactions (eg thermal decomposition) are endothermic (take in energy from the surroundings).
- Reversible reactions are endothermic in one direction and exothermic in the other.
- Acids have a pH of less than 7 and are neutralised by bases.
- Metal oxides or hydroxides are bases; ammonia is also a base.
- Alkalis are bases dissolved in water. They have a pH of greater than 7.
- Acids contain $H^+$ ions, alkalis contain $OH^-$ ions. These react to form water.
- Salts form when acids react with bases, alkalis, or reactive metals.
- Soluble salts are prepared by neutralisation followed by crystallisation.
- Insoluble salts are made by precipitation reactions.
- Precipitation is used in water treatment.
- In electrolysis an electric current is passed through an ionic compound (molten or in solution). The compound breaks down into elements.
- Reduction happens at the negative electrode. Positive ions gain electrons.
- Oxidation happens at the positive electrode. Negative ions lose electrons.
- Metal objects are electroplated by making them a negative electrode and placing them in a solution of suitable ions.
- Aluminium is extracted by electrolysis of molten bauxite dissolved in cryolite.
- Electrolysis of brine produces chlorine, sodium hydroxide, and hydrogen.

combustion

mass of product formed per second

RATE OF CHEMICAL REACTIONS

reduce cost of industrial reactions

using catalys

alters rate but unchanged at

hydrogen, sodium hydroxide, and chlorine

sodium chloride

breaking down ionic compoun

ELECTROLYSIS

electroplating

of negative electrode dipped in silver ions, chronium ions etc

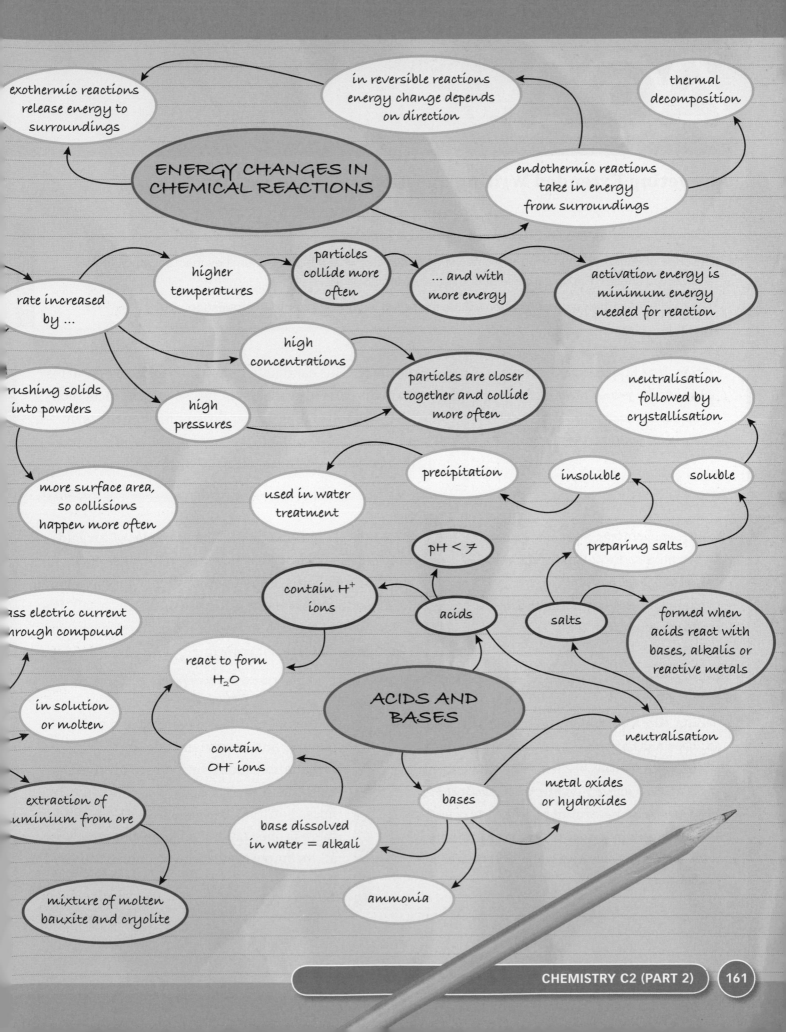

exothermic reactions release energy to surroundings

in reversible reactions energy change depends on direction

thermal decomposition

**ENERGY CHANGES IN CHEMICAL REACTIONS**

endothermic reactions take in energy from surroundings

rate increased by ...

higher temperatures

particles collide more often

... and with more energy

activation energy is minimum energy needed for reaction

high concentrations

crushing solids into powders

high pressures

particles are closer together and collide more often

neutralisation followed by crystallisation

more surface area, so collisions happen more often

used in water treatment

precipitation

insoluble

soluble

pH < 7

preparing salts

contain H$^+$ ions

ass electric current through compound

acids

salts

formed when acids react with bases, alkalis or reactive metals

react to form H$_2$O

in solution or molten

**ACIDS AND BASES**

neutralisation

contain OH$^-$ ions

metal oxides or hydroxides

extraction of aluminium from ore

bases

base dissolved in water = alkali

mixture of molten bauxite and cryolite

ammonia

# Answering Extended Writing questions

**QUESTION**

Acids and alkalis are two important types of chemical substance.

Describe the difference between these two substances and what happens when they react together. In your answer you should give an example of both an acid and an alkali.

**The quality of written communication will be assessed in your answer to this question.**

**G–E**

Acids, like hydrolic acid have a PH of below 7 and alkalis above 7. They cancel each other out when they react to make PH7.

**Examiner:** This candidate knows how to recognise acids and alkalis from pH values (but has shown pH incorrectly as 'PH'). However, the answer doesn't include the term 'neutralise'. The candidate has not named an alkali and isn't very close with the spelling of 'hydrochloric acid'.

**D–C**

Sulfuric acid has a pH of 1 and will newtralise sodium hydroxide which is an alkali and has a pH of over 7. When they react they make a salt, like sodium chloride.

**Examiner:** This candidate knows names of an acid and alkali and describes how they neutralise each other (though this is spelt wrongly). The answer should also mention that water is formed as well as a salt. It's a shame that the salt formed in the reaction discussed (sodium sulfate) is not named.

**B–A\***

Acids, like hydrochloric acid, have H+ ions and so have a pH of below 7. Alkalis are solubull bases, like sodium hydroxide and have OH ions and a pH of above 7. When they react the H+ and OH ions neutralise to make water and a salt (sodium chloride) is also made. The pH is now 7 because salt and water are neutral substances.

**Examiner:** This is a detailed and well-planned answer and includes an explanation about the ions are involved in neutralisation reactions. The candidate has missed off the charge on the OH⁻ ion, and there is one spelling mistake, but otherwise all the important scientific information is here.

# Exam-style questions

**1** Olivia heated 200 g of water using 1 g of two fuels, ethanol and hexane, in an experiment to help her decide which one is the best fuel.

| Fuel | ethanol | hexane |
|---|---|---|
| Start temperature (°C) | 21 | 21 |
| End temperature (°C) | 43 | |
| Temperature rise (°C) | | 16 |

A03 **a** Complete the missing boxes.

A03 **b** Choose the correct description of the energy changes:

    **i** Both reactions give out energy.

    **ii** Hexane gives out energy and ethanol takes in energy.

    **iii** Hexane takes in energy and ethanol gives out energy.

A03 **c** Which fuel is the best?

A02 **d** Give one way in which Olivia made her experiment a fair test.

**2** Jack investigated the factors affecting the reaction between magnesium and hydrochloric acid. He performed the reaction first with a low concentration of hydrochloric acid, and then with a high concentration.

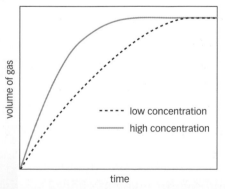

A03 **a** Which reaction has the fastest rate?

A03 **b** What can Jack conclude about the effect of increasing concentration on rate of reaction?

A02 **c** In another experiment, Jack found that increasing the temperature made the reaction faster. Use ideas about particles to explain why.

**3** Hydrochloric acid can be neutralised by a solution of calcium hydroxide. A salt and water are formed.

A02 **a** Name the salt that is formed.

A01 **b** What name is given to soluble bases such as calcium hydroxide?

A02 **c** Write a balanced symbol equation to show how the hydroxide ions in calcium hydroxide are neutralised by ions from the hydrochloric acid.

## Extended Writing

**4** Catalysts are often used in industrial reactions. What are catalysts and why are they important in industry? Give an example of their use in industry.

A01

**5** Shafiq wants to make a sample of solid magnesium sulfate, a soluble salt, from solid magnesium oxide and sulfuric acid. Write a set of instructions that he can follow to make this salt.

A02

**6** Electrolysis of brine (sodium chloride) is used to make hydrogen, sodium hydroxide, and chlorine. Describe what happens in this process and why it is important.

A01

D–C

G–E

B–A*

G–E

D–C

B–A*

G–E

D–C

# C3 Part 1

# Water, energy, and the periodic table

## Why study this unit?

Finding patterns and making predictions are key to science. Dmitri Mendeleev did just that when, in 1869, he created the first periodic table of the elements.

Clean drinking water is vital for life. Chemists are at the centre of the processes that make our water safe to drink.

Many people love their cars. But what will happen when fossil fuels run out? Chemists are working to bring hydrogen to our fuel tanks, creating a more sustainable transport future.

In this unit you will learn about the patterns in the periodic table, and how they help us to predict properties. You will discover how water safety and quality are ensured. You will also learn about energy changes in chemical reactions, and evaluate a new vehicle fuel – hydrogen.

## You should remember

1  Everything is made up of atoms of about a hundred elements, listed in the periodic table.

2  Metals (on the left of the periodic table) are usually shiny, and good conductors of heat and electricity.

3  Non-metals (on the right of the periodic table) are not shiny, and most are poor conductors of heat and electricity.

4  Filtration is used to separate a mixture of a solid and a liquid. Distillation can be used to separate mixtures of liquids, or a liquid from a solution.

5  Exothermic reactions (such as combustion reactions) transfer energy to the surroundings, or give out energy.

6  Endothermic reactions take in energy from the surroundings.

7  Burning fossil fuels produces pollutants.

This is a cholera bacterium. Its flagellum (tail) propels it through water, and into the body of the next victim of the disease. Cholera causes severe diarrhoea and vomiting, leading to dehydration and death.

In the nineteenth century, millions died of the disease all over the world. Today, the addition of chlorine to drinking water has virtually eliminated the disease in richer countries. But cholera epidemics continue to claim the lives of thousands in poorer countries.

▲ Fashionable women of the 1860s

▲ John Newlands was chief chemist in a London sugar factory

## Decade of discovery

In Britain, fashionable women dressed like those shown on the left. In America, slavery was abolished. In France, the first true bicycles were invented. The decade? The 1860s.

The 1860s were important in chemistry, too. By then, chemists knew of more than 50 elements. The Italian Stanislao Cannizzaro had worked out their atomic weights. Chemists puzzled over properties, looking for patterns. Was there a link between an element's properties and its atomic weight? How could elements best be classified into groups?

## Newlands' octaves

In the mid 1860s, John Newlands made progress in grouping the elements. He listed the 56 elements then known in order of increasing atomic weight. There was a pattern – every eighth element had similar properties. Newlands used this pattern to group the elements. He called his discovery the **law of octaves**, after the musical scale.

Newlands' grouping was not perfect. At a meeting of the London Chemical Society, chemists criticised his law of octaves. They asked why copper was grouped with lithium, sodium, and potassium, when its properties were so different. And did it make sense to include the metal nickel in the same group as fluorine, chlorine, and bromine?

> **A** What information, supplied by Cannizzaro, helped Newlands to come up with his law of octaves?
>
> **B** What problem did other chemists notice with Newlands' law of octaves?

## Mendeleev's masterpiece

In 1869, a Russian chemistry professor made a vital breakthrough. On 1 March, Dmitri Mendeleev had planned to visit a cheese factory. But the weather was terrible, so he decided to work from home.

There, he made lots of small cards. On each, he wrote the name of an element, its properties, and its atomic weight. He tried placing the cards in different patterns.

By lunchtime, Mendeleev had come up with an arrangement that worked. The elements were in order of increasing atomic mass. At the same time, elements with similar properties were grouped together. Mendeleev wrote this arrangement on the back of an envelope. It was the first **periodic table**.

## Filling in the gaps

Mendeleev knew he had discovered something important. But he realised that his periodic table was not perfect. For example, when placed in atomic weight order, iodine and tellurium seemed to be in the wrong groups. So Mendeleev swapped their positions. Iodine was now in a group with fluorine, chlorine, and bromine. All four of these elements have similar properties.

Mendeleev studied the patterns in his periodic table. He realised that it did not include all the elements. Had some elements not yet been discovered? Mendeleev left gaps for the missing elements, and predicted their properties. Over the next few years, other chemists searched for elements to fill the gaps:

- In 1875, Frenchman Paul-Émile Lecoq de Boisbaudran discovered the element under aluminium. He called it gallium. Its properties were those predicted by Mendeleev.
- In 1879, the Swede Lars Nilson found another missing element. He called it scandium. Again, its properties matched those predicted by Mendeleev.

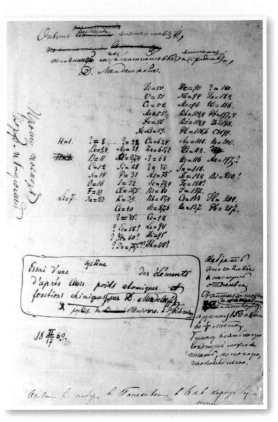

▲ The first periodic table

### Did you know...?

The periodic table is so named because similar properties occur at regular intervals – just like the repeating pattern of menstrual periods.

### Exam tip

✔ In modern periodic tables, elements with similar properties are in the same vertical column – or group – of the periodic table.

### Key words

law of octaves, periodic table

### Questions

1 Name the chemist who discovered the law of octaves.

2 Copy and complete: Mendeleev's periodic table listed the elements in order of increasing atomic _____.

3 In the periodic table, what is a group?

4 Why is the periodic table called the periodic table?

5 List two problems Mendeleev noticed with his periodic table. Describe what he did to address these problems.

## Learning objectives

After studying this topic, you should be able to:

✔ explain the links between an element's position in the periodic table, its atomic number, and its electron configuration

▲ The Wright brothers' aeroplane

---

**A** Name the sub-atomic particle discovered by J. J. Thomson.

**B** Describe Rutherford's two linked discoveries.

---

## Tiny particles, big discoveries

The twentieth century got off to a flying start. In 1902, the Wright brothers made the first controlled, powered, heavier than air flight. In 1903, the Ford Car Company produced its first car.

The new century saw breakthroughs in chemistry, too. In 1897, Joseph John Thomson had experimented with cathode ray tubes and discovered electrons. In 1909, Ernest Rutherford bombarded gold foil with radioactive particles, and discovered that most of the mass of an atom is concentrated in a tiny, positively charged nucleus.

In 1911, Antonius van den Broek suggested that the amount of positive charge on the nucleus of an atom of an element might be linked to its position in the periodic table. This idea inspired British student Henry Moseley. Moseley carried out experiments that provided practical evidence to support van den Broek's theory.

In 1919, Rutherford made another vital breakthrough. He discovered the sub-atomic particle that gives nuclei their positive charge. He named this particle the proton.

## Atomic number

The number of protons in an atom of an element is the **atomic number** of the element.

▼ The modern periodic table

| 1 | 2 | | | | | | | | | | 3 | 4 | 5 | 6 | 7 | 0 |
|---|---|---|---|---|---|---|---|---|---|---|---|---|---|---|---|---|
| | | | | | **Key** | | 1<br>**H**<br>Hydrogen<br>1 | | | | | | | | | 4<br>**He**<br>Helium<br>2 |
| | | | | | relative atomic mass<br>**atomic symbol**<br>name<br>atomic (proton) number | | | | | | | | | | | |
| 7<br>**Li**<br>Lithium<br>3 | 9<br>**Be**<br>Beryllium<br>4 | | | | | | | | | | 11<br>**B**<br>Boron<br>5 | 12<br>**C**<br>Carbon<br>6 | 14<br>**N**<br>Nitrogen<br>7 | 16<br>**O**<br>Oxygen<br>8 | 19<br>**F**<br>Fluorine<br>9 | 20<br>**Ne**<br>Neon<br>10 |
| 23<br>**Na**<br>Sodium<br>11 | 24<br>**Mg**<br>Magnesium<br>12 | | | | | | | | | | 27<br>**Al**<br>Aluminium<br>13 | 28<br>**Si**<br>Silicon<br>14 | 31<br>**P**<br>Phosphorus<br>15 | 32<br>**S**<br>Sulfur<br>16 | 35.5<br>**Cl**<br>Chlorine<br>17 | 40<br>**Ar**<br>Argon<br>18 |
| 39<br>**K**<br>Potassium<br>19 | 40<br>**Ca**<br>Calcium<br>20 | 45<br>**Sc**<br>Scandium<br>21 | 48<br>**Ti**<br>Titanium<br>22 | 51<br>**V**<br>Vanadium<br>23 | 52<br>**Cr**<br>Chromium<br>24 | 55<br>**Mn**<br>Manganese<br>25 | 56<br>**Fe**<br>Iron<br>26 | 59<br>**Co**<br>Cobalt<br>27 | 59<br>**Ni**<br>Nickel<br>28 | 63.5<br>**Cu**<br>Copper<br>29 | 65<br>**Zn**<br>Zinc<br>30 | 70<br>**Ga**<br>Gallium<br>31 | 73<br>**Ge**<br>Germanium<br>32 | 75<br>**As**<br>Arsenic<br>33 | 79<br>**Se**<br>Selenium<br>34 | 80<br>**Br**<br>Bromine<br>35 | 84<br>**Kr**<br>Krypton<br>36 |
| 85<br>**Rb**<br>Rubidium<br>37 | 88<br>**Sr**<br>Strontium<br>38 | 89<br>**Y**<br>Yttrium<br>39 | 91<br>**Zr**<br>Zirconium<br>40 | 93<br>**Nb**<br>Niobium<br>41 | 96<br>**Mo**<br>Molybdenum<br>42 | [98]<br>**Tc**<br>Technetium<br>43 | 101<br>**Ru**<br>Ruthenium<br>44 | 103<br>**Rh**<br>Rhodium<br>45 | 106<br>**Pd**<br>Palladium<br>46 | 108<br>**Ag**<br>Silver<br>47 | 112<br>**Cd**<br>Cadmium<br>48 | 115<br>**In**<br>Indium<br>49 | 119<br>**Sn**<br>Tin<br>50 | 122<br>**Sb**<br>Antimony<br>51 | 128<br>**Te**<br>Tellurium<br>52 | 127<br>**I**<br>Iodine<br>53 | 131<br>**Xe**<br>Xenon<br>54 |
| 133<br>**Cs**<br>Caesium<br>55 | 137<br>**Ba**<br>Barium<br>56 | 139<br>**La***<br>Lanthanum<br>57 | 178<br>**Hf**<br>Hafnium<br>72 | 181<br>**Ta**<br>Tantalum<br>73 | 184<br>**W**<br>Tungsten<br>74 | 186<br>**Re**<br>Rhenium<br>75 | 190<br>**Os**<br>Osmium<br>76 | 192<br>**Ir**<br>Iridium<br>77 | 195<br>**Pt**<br>Platinum<br>78 | 197<br>**Au**<br>Gold<br>79 | 201<br>**Hg**<br>Mercury<br>80 | 204<br>**Tl**<br>Thallium<br>81 | 207<br>**Pb**<br>Lead<br>82 | 209<br>**Bi**<br>Bismuth<br>83 | [209]<br>**Po**<br>Polonium<br>84 | [210]<br>**At**<br>Astatine<br>85 | [222]<br>**Rn**<br>Radon<br>86 |
| [223]<br>**Fr**<br>Francium<br>87 | [226]<br>**Ra**<br>Radium<br>88 | [227]<br>**Ac***<br>Actinium<br>89 | [261]<br>**Rf**<br>Rutherfordium<br>104 | [262]<br>**Db**<br>Dubnium<br>105 | [266]<br>**Sg**<br>Seaborgium<br>106 | [264]<br>**Bh**<br>Bohrium<br>107 | [277]<br>**Hs**<br>Hassium<br>108 | [268]<br>**Mt**<br>Meitnerium<br>109 | [271]<br>**Ds**<br>Darmstadtium<br>110 | [272]<br>**Rg**<br>Roentgenium<br>111 | | | | | | | |

Elements with atomic numbers 112–116 have been reported but not fully authenticated

\* The Lanthanides (atomic numbers 58–71) and the Actinides (atomic numbers 90–103) have been omitted.
**Cu** and **Cl** have not been rounded to the nearest whole number

Having discovered the proton, chemists tried arranging the elements in order of increasing atomic number. This placed all elements in appropriate groups. The problems of Mendeleev's periodic table, based on atomic weights, were solved.

## Electrons matter too

Modern periodic tables still arrange the elements in order of increasing atomic number. The positions of the elements are also linked to their electronic structures. Elements in the same group have the same number of electrons in their highest occupied energy level. For example, the elements in Group 2 have the electronic structures shown below. There are two electrons in the highest occupied energy level of the atoms of each element.

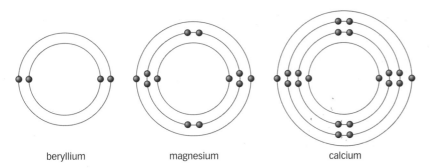

beryllium      magnesium      calcium

▲ The Group 2 elements all have two electrons in their highest occupied energy level

The periodic table started as a scientific curiosity. It became a useful tool. Today, chemists view it as an important summary of the structure of atoms.

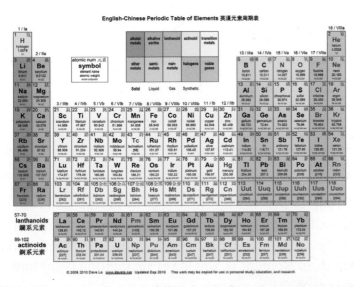

▲ Chemists all over the world use the periodic table. This one is from China.

**Key words**

atomic number

## Exam tip

✔ The number of electrons in the highest occupied energy level for the elements in the main groups of the periodic table is equal to the group number.

## Questions

1. Give the meaning of the term atomic number.

2. Copy and complete: Modern periodic tables arrange the elements in order of increasing atomic _____.

3. Name the first three elements in Group 1 of the periodic table.

4. Draw the electronic structures of the elements lithium, sodium, and potassium. How many electrons are in the highest occupied energy level of these elements?

5. Predict the number of electrons in the highest occupied energy level of the elements fluorine, chlorine, and bromine.

E

C

A*

## Key words

alkali metal

▲ Bolivia has huge reserves of lithium in salt flats like these

## Lovely lithium

Lithium is in demand. Lithium batteries power mobile phones, personal music players, and artificial heart pacemakers. Electric cars have lithium batteries, too. As more and more people buy electric cars, so the demand for lithium will increase.

The South American country of Bolivia has huge reserves of lithium. In 2010, the Bolivian government was testing out ways of getting lithium from its vast salt flats. The government wants Bolivians, not foreign companies, to benefit from selling lithium.

## Soft, light, and low

Group 1
the alkali metals

▲ Group 1 elements are on the left of the periodic table

Lithium is in Group 1 of the periodic table, called the **alkali metals**. Lithium has the lowest density of all the metals. The other Group 1 elements also have low densities.

| Element | Density (g/cm³) |
|---|---|
| lithium | 0.53 |
| sodium | 0.97 |
| potassium | 0.86 |
| rubidium | 1.53 |

The Group 1 elements are very soft – you can easily cut them with a knife. They have lower melting points and boiling points than most other metals. In Group 1, the further down the group an element is, the lower its melting point and boiling point.

A  The density of water is 1.00 g/cm³. Name the alkali metals that float on water.

B  Draw a bar chart to show the densities of the alkali metals. Describe the overall pattern in density. Predict the density of caesium.

# Reactions with non-metals

The alkali metals react vigorously with chlorine. Sodium, for example, burns with a bright orange flame in the pale green gas. The product is a white solid, sodium chloride.

$$\text{sodium} + \text{chlorine} \rightarrow \text{sodium chloride}$$
$$2Na(s) + Cl_2(g) \rightarrow 2NaCl(s)$$

Potassium reacts with chlorine in a similar way. So does lithium.

Potassium chloride, sodium chloride, and lithium chloride are white solids. They dissolve in water to form colourless solutions.

The alkali metal chlorides are ionic compounds. The atoms of all Group 1 elements have one electron in their highest occupied energy level. So when potassium reacts with chlorine, each potassium atom transfers one electron to a chlorine atom. This forms positive potassium ions, $K^+$, and negative chloride ions, $Cl^-$. Both types of ion have eight electrons in their highest occupied energy level.

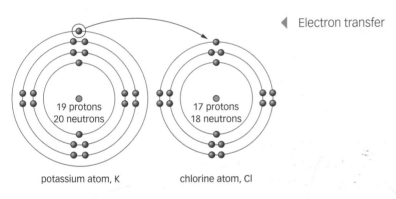

◀ Electron transfer

potassium atom, K          chlorine atom, Cl

▲ Potassium burns vigorously in chlorine

## Questions

1 Describe three properties of the alkali metals.

2 Name the alkali metal with the lowest density.

3 Write a word equation for the reaction of potassium with chlorine.

4 Name the type of bonding in lithium chloride. Give one property of this solid.

5 Lithium reacts with bromine to form lithium bromide. Write a balanced symbol equation for the reaction.

### Did you know...?

Lithium chloride can absorb huge amounts of water. It is used to dry the gases in submarine air conditioners.

### Exam tip          AQA

✔ Remember, alkali metals react with chlorine to form ionic compounds in which the metal ion has a charge of +1.

### Learning objectives

After studying this topic, you should be able to:

✔ describe the reactions of the alkali metals with oxygen and water

✔ explain the trend in reactivity of the alkali metals

▲ Physicists adjusting the settings of an atomic clock

### Did you know...?

Streams of caesium ions ejected through the nozzle of a 'thruster' are used to steer satellites.

**A** Name the product of the reaction of lithium with oxygen.

**B** Write a symbol equation for the reaction of potassium with oxygen.

## Keeping time

Does your watch, computer, or mobile phone keep accurate time? It will if it includes a radio clock. In the UK, radio clocks receive signals from three atomic clocks in Cumbria. Atomic clocks are based on electron movements in caesium atoms. They can be accurate to 1 second in 30 million years.

Caesium fluoride and caesium iodide absorb gamma rays (from radioactive material) and X-rays. As they absorb this radiation, they give out light. This property makes caesium fluoride and caesium iodide useful for monitoring radiation.

Caesium is in Group 1 of the periodic table, along with lithium, sodium, potassium, and rubidium. It is an alkali metal.

## Reactions with oxygen

We saw on the previous spread that the alkali metals react vigorously with chlorine. They also react quickly with oxygen, another non-metal. At room temperature, their surfaces quickly tarnish when exposed to air. This is why they are stored under oil or grease.

The alkali metals burn in air and oxygen. For example:

$$\text{sodium} + \text{oxygen} \rightarrow \text{sodium oxide}$$
$$4Na(s) + O_2(g) \rightarrow 2Na_2O(s)$$

◀ Sodium burns brightly in oxygen

Sodium oxide is a white solid. It dissolves in water to make a colourless solution.

Alkali metal oxides are ionic compounds. They are made up of:

- oxide ions, $O^{2-}$
- metal ions with a charge of +1, for example $Na^+$ or $Li^+$.

## Reactions with water

The alkali metals react vigorously with water. As they react, they zoom around on the water surface, propelled by the bubbles of hydrogen gas produced in the reaction. The other product of these reactions is a metal hydroxide. This dissolves in water to give an alkaline solution. For example:

sodium + water → sodium hydroxide + hydrogen

$$2Na(s) + 2H_2O(l) → 2NaOH(aq) + H_2(g)$$

## Group trend

Alkali metal reactions get more vigorous going down the group.

Going down the group, alkali metal reactions get more vigorous because the electrons involved in the reaction are further from the nucleus. This makes them less strongly attracted to the nucleus. So, in a reaction, an atom of an element at the bottom of the group gives away an electron more easily than an atom of an element at the top of the group.

▲ Sodium reacts vigorously with water

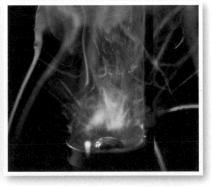

▲ Potassium is below sodium in Group 1. It reacts more vigorously with water than sodium does.

◀ Caesium is below both sodium and potassium in Group 1. It reacts violently with water.

### Questions

1 Name the product of the reaction of sodium with oxygen.

2 Name the products of the reaction of potassium with water.

3 Write a word equation for the reaction of caesium with water.

4 Suggest why the Group 1 elements are called the alkali metals.

5 Write a balanced symbol equation for the reaction of rubidium (Rb) with water.

6 Describe and explain the trend in reactivity of the alkali metals.

E
C
A*

## Catalysts, coins, and batteries

What links the pictures?

The answer is nickel. The coin is an alloy of copper and nickel. Nickel catalyses the reaction of hydrogen with vegetable oils to make margarine. Nickel–cadmium batteries can be recharged more than a thousand times.

## Physical properties

Nickel is in the central block of the periodic table. It is a **transition element**. The transition elements are metals. Their properties have some similarities to those of the alkali metals. They conduct electricity, for example, and have shiny surfaces when freshly cut. But there are differences, too. Compared to the alkali metals, the transition elements

* are stronger and harder
* have higher densities
* have higher melting points (except for mercury, which is liquid at room temperature).

| Name of metal | Melting point (°C) | Density (g/cm³) |
|---|---|---|
| lithium | 180 | 0.53 |
| sodium | 98 | 0.97 |
| nickel | 1453 | 8.9 |
| palladium | 1550 | 12.0 |
| X | 1769 | 21.4 |

**A** Describe three differences between a typical alkali metal and a typical transition element.

**B** Predict whether metal X in the table is an alkali metal or a transition element. Give a reason for your decision.

# Chemical reactions

The transition elements are less reactive than the alkali metals. For example, at room temperature the alkali metals react quickly with water and oxygen. The transition elements react slowly, if at all:

- Platinum and gold do not react with water or oxygen – that's why they make good jewellery, and why gold is used for electrical connections.
- Iron reacts with water and oxygen at room temperature, but slowly. The product is hydrated iron oxide, or rust.

## Colours, catalysts, and ions

Many transition elements form more than one type of ion. For example, iron has two main oxides, $FeO$ and $Fe_2O_3$. The oxide ion has a charge of $-2$ in both oxides. So $FeO$ includes $Fe^{2+}$ ions and $Fe_2O_3$ includes $Fe^{3+}$ ions.

Many transition elements form coloured compounds.

▲ Iron(II) oxide, FeO, is green

▲ Iron(III) oxide, $Fe_2O_3$, is brown

◄ Vanadium compounds come in several colours. The colour of a compound depends on the charge of its vanadium ion.

Transition elements are important catalysts. For example, in catalytic convertors, platinum, palladium, and rhodium convert dangerous exhaust gases to ones that are less harmful.

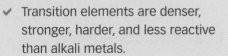

▲ Compounds containing $Cu^{2+}$ ions are blue or green

## Questions

1. Describe one way in which transition metals are used.
2. Describe one difference between the reactions of alkali metals and the reactions of transition metals. ↓E
3. Explain why different vanadium compounds come in different colours. ↓C
4. Work out the charge on the vanadium ion in the compound $V_2O_5$. ↓A*

**A** Describe the pattern in boiling points as you go down the halogen group.

**B** Give the states of bromine and iodine at room temperature (20 °C).

◀ Chlorine reacts with iron to make iron(III) chloride

## Deadly … but vital

Group 7 is home to five deadly non-metal elements, the **halogens**.

| Group | | | | | | | | | | | | | | | | | Group 7 the halogens |
|---|---|---|---|---|---|---|---|---|---|---|---|---|---|---|---|---|---|
| | | | | | | | H | | | | | | | | | | He |
| Li | Be | | | | | | | | | | | B | C | N | O | F | Ne |
| Na | Mg | | | | | | | | | | | Al | Si | P | S | Cl | Ar |
| K | Ca | Sc | Ti | V | Cr | Mn | Fe | Co | Ni | Cu | Zn | Ga | Ge | As | Se | Br | Kr |
| Rb | Sr | Y | Zr | Nb | Mo | Tc | Ru | Rh | Pd | Ag | Cd | In | Sn | Sb | Te | I | Xe |
| Cs | Ba | La | Hf | Ta | W | Re | Os | Ir | Pt | Au | Hg | Tl | Pb | Bi | Po | At | Rn |
| Fr | Ra | Ac | Rf | Db | Sg | Bh | Hs | Mt | Ds | Rg | | | | | | | |

▲ Group 7 elements are towards the right of the periodic table

But the halogens are not all bad. Chlorine destroys bacteria and viruses, so it is added to water to make it safe to drink. Fluoride compounds strengthen teeth. Iodine-containing hormones are vital for normal growth and development. They also keep your body temperature constant.

## Physical properties

The halogens exist as two-atom molecules, such as $F_2$ and $Br_2$. The table shows some of their properties.

| Element | Colour | Melting point (°C) | Boiling point (°C) |
|---|---|---|---|
| fluorine | pale yellow | –220 | –188 |
| chlorine | green | –101 | –34.7 |
| bromine | orange/brown | –7.2 | 58.8 |
| iodine | grey/black with violet vapour | 114 | 184 |

## Comparing reactivity

A teacher reacts chlorine gas with iron wool.

The reaction is fast and fierce. Reddish-brown iron(III) chloride forms.

$$\text{iron} + \text{chlorine} \rightarrow \text{iron(III) chloride}$$
$$2Fe(s) + 3Cl_2(g) \rightarrow 2FeCl_3(s)$$

The other halogens also react with iron:

- Bromine forms iron(III) bromide. The reaction is slower than that of chlorine with iron.
- Iodine forms iron(III) iodide. The reaction is even slower than that of bromide with iron.

These reactions show that the halogens get less reactive as you go down the group.

## Explaining reactivity

In the reactions of iron with the halogens, iron atoms give electrons to halogen atoms. This forms **halide ions**, such as $Cl^-$ and $Br^-$. The smaller atoms at the top of the group gain electrons more easily. This is because the negatively charged electrons are added to an energy level that is closer to the positively charged nucleus. So the attraction between the nucleus and electrons is stronger. Overall, the higher the energy level, the less easily electrons are gained.

## Displacement reactions

You can compare the reactivity of halogens in **displacement reactions**.

Miranda adds a solution of chlorine gas to potassium bromide solution. Yellow-orange bromine forms. Chlorine has 'pushed', or displaced, the bromide ion out of its compound. This shows that chlorine is more reactive than bromine.

$$\text{chlorine} + \begin{array}{c}\text{potassium}\\\text{bromide}\end{array} \rightarrow \begin{array}{c}\text{potassium}\\\text{chloride}\end{array} + \text{bromine}$$

$$Cl_2(aq) + 2KBr(aq) \rightarrow 2KCl(aq) + Br_2(aq)$$

Chlorine is also more reactive than iodine. So chlorine displaces iodine from its compounds.

$$\text{chlorine} + \begin{array}{c}\text{potassium}\\\text{iodide}\end{array} \rightarrow \begin{array}{c}\text{potassium}\\\text{chloride}\end{array} + \text{iodine}$$

$$Cl_2(aq) + 2KI(aq) \rightarrow 2KCl(aq) + I_2(aq)$$

Bromine displaces iodine from its compounds, too.

$$Br_2(aq) + 2KI(aq) \rightarrow 2KBr(aq) + I_2(aq)$$

## Questions

1. Describe the pattern in melting points as you go down the group of halogens.

2. Give the formulae of fluorine gas and chlorine gas.

3. Write a word equation for the displacement reaction of bromine with potassium iodide.

4. Explain why the halogens get less reactive going down the group.

5. Miranda adds bromine solution to potassium chloride solution. Explain why there is no reaction.

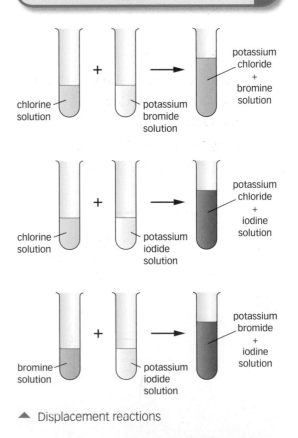

▲ Displacement reactions

## Learning objectives

After studying this topic, you should be able to:

✓ explain what makes water hard or soft

✓ describe how to measure hardness

## Key words

**hard water, soft water, permanent hard water, temporary hard water**

▲ Getting used to soft water

## Did you know...?

Tea made from hard water tastes different from the same tea made with soft water. Some British tea companies blend teas especially for hard water areas.

## New water

Ben moves from Wiltshire to Cheshire. He quickly settles into his new home and new school. But nothing will persuade him to drink the water in Cheshire. 'It tastes disgusting,' he says. Ben's neighbour can't see the problem. 'I've been drinking it all my life,' she says. 'It's delicious!' What makes the water in different areas taste so different?

## Hard or soft?

The water Ben drank in Wiltshire is **hard water**. Hard water contains dissolved compounds, usually of calcium or magnesium. Water becomes hard when it flows through chalk or limestone rocks. Here's how: *How does water flow through chalk or limestone rocks*

- As rain falls, carbon dioxide from the air dissolves in it. Carbonic acid forms. So rainwater is weakly acidic.

$$\text{water} + \text{carbon dioxide} \rightarrow \text{carbonic acid}$$
$$H_2O(l) + CO_2(g) \rightarrow H_2CO_3(aq)$$

- As rainwater flows through chalk or limestone, its carbonic acid reacts with calcium carbonate in the rock. The product of the reaction is calcium hydrogencarbonate. Calcium hydrogencarbonate is soluble in water. Its calcium ions make water hard.

$$\begin{array}{ccc} \text{carbonic} & \text{calcium} & \text{calcium} \\ \text{acid} & + \ \text{carbonate} & \rightarrow \ \text{hydrogencarbonate} \end{array}$$
$$H_2CO_3(aq) + CaCO_3(s) \rightarrow Ca(HCO_3)_2(aq)$$

▲ Water from chalky areas is hard

Gypsum rock (mainly calcium sulfate) also makes water hard. When a river flows over gypsum, calcium sulfate dissolves in the water. The water now contains dissolved calcium ions.

**Soft water** does not contain dissolved calcium or magnesium ions.

# Measuring hardness

Soft water lathers easily with soap. Calcium and magnesium ions in hard water react with soap to form scum. So hard water needs more soap to form a lather.

Samia wants to compare the hardness of different water samples. She adds soap solution to 10 cm³ of each water sample. The more soap solution needed to form a permanent lather, the harder the water. The table shows her results.

| Water sample | Volume of soap solution needed to make a lather (cm³) | Volume of soapless detergent needed to make a lather (cm³) |
|---|---|---|
| rainwater | 1 | 1 |
| tap water from town A | 1 | 1 |
| tap water from town B | 9 | 1 |

# Hard forever?

There are two types of hard water:

- **Permanent hard water** remains hard, even when it is boiled. Water that has flowed over gypsum is permanent hard water.
- **Temporary hard water** is softened when it boils. Water that has flowed over chalk or limestone is temporary hard water.

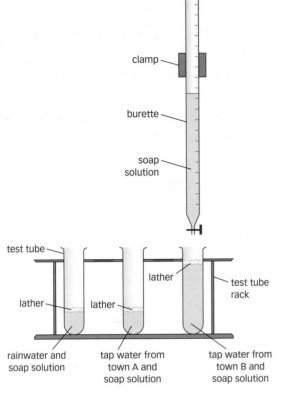

▲ Samia adds soap solution from the burette to test the hardness of the water samples

**A** Which town has hard water?

**B** Is rainwater hard or soft?

## Questions

1. What is hard water?
2. Name three types of rock that make water hard.
3. Explain how water becomes hard. Include an equation to help you explain.
4. Describe the difference between temporary hard water and permanent hard water.
5. Use the data in the table to predict whether or not soapless detergents react with hard water to form scum. Explain your answer.

↓ E
↓ C
↓ A*

**Exam tip**

✔ Make sure you know the difference between temporary hard water and permanent hard water.

## Hard or soft – pros and cons

I hate our hard water. It costs me a fortune in soap and washing powder.

Hard water is a nuisance. It makes scale in our kettles and central heating boiler. Scale makes our electric kettle and gas boiler less efficient than they should be. My electricity and gas bills are huge.

We have hard water. Its calcium compounds help my children's bones and teeth to grow. It tastes great too.

I read an article about hard and soft water in a scientific journal. The ions dissolved in hard water may help to reduce the chance of getting heart disease.

## Softening hard water

There is money to be made in water softening. Many companies supply water-softening materials and equipment to UK homes and factories. All softening methods have the same aim – to remove the dissolved calcium ions ($Ca^{2+}$) and magnesium ions ($Mg^{2+}$) that make water hard.

### Washing soda

A cheap and simple way of softening hard water is to add sodium carbonate, or **washing soda**. Most washing powders include washing soda.

Sodium carbonate is soluble in water. When you add it to hard water, its carbonate ions react with dissolved calcium and magnesium ions. Calcium carbonate and magnesium carbonate form as precipitates.

**A** Give the chemical name for washing soda.

**B** Explain how washing soda softens water.

The ionic equation below summarises the reaction that removes calcium ions. Only the ions that take part in the reaction are included.

calcium ions + carbonate ions → calcium carbonate

$$Ca^{2+}(aq) + CO_3^{2-}(aq) \rightarrow CaCO_3(s)$$

## Ion exchange columns

Some people buy **ion exchange columns** to soften their water. Ion exchange columns swap calcium and magnesium ions from hard water with sodium or hydrogen ions.

In this ion exchange resin, sodium ions are attached to the resin. Hard water flows in at the top. It trickles through the resin. On its way down, its calcium and magnesium ions swap with sodium ions and stick to the resin. The water that comes out of the bottom has sodium ions dissolved in it. It is no longer hard.

After a while, the column becomes saturated with calcium and magnesium ions. It no longer works. You need to pour sodium chloride solution through the column to flush out the calcium and magnesium ions and replace them with sodium ions. The column is now ready to use again.

An ion exchange column can supply a whole house with softened water. Scale no longer forms in kettles and boilers, so energy costs are reduced.

However, ion exchange columns increase the amount of sodium in the water. Dissolved sodium is not good for heart health, or for babies. So it is best not to use softened water for drinking, cooking, or babies' bottles.

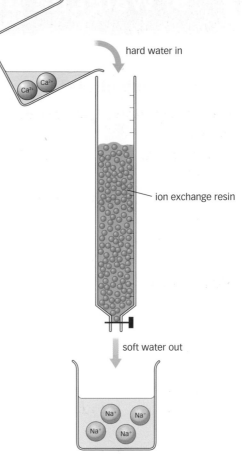

▲ An ion exchange column. Ions are not drawn to scale.

### Removing temporary hardness

You can remove temporary hardness by boiling. The dissolved hydrogencarbonate ions decompose on heating to produce carbonate ions. The carbonate ions react with calcium or magnesium ions in the water to form a precipitate. The precipitate is the scale you see in kettles. Of course, it does not make economic sense to soften water by boiling on a large scale.

calcium → calcium + carbon + water
hydrogencarbonate   carbonate   dioxide

$$Ca(HCO_3)_2(aq) \rightarrow CaCO_3(s) + CO_2(g) + H_2O(l)$$

## Questions

1  Describe two problems of hard water.

2  Describe two benefits of hard water.

3  Describe an economic benefit of softening hard water.

4  Write a balanced ionic equation to summarise how washing soda removes dissolved magnesium ions from solution.

5  Explain how an ion exchange resin makes hard water soft.

▲ Baths use lots of water

**A** Explain what makes UK tap water safe to drink.

**B** How much more water does an average European person use than an average person in a poorer part of the world?

 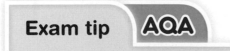
## How much?

How much water do you use each day? If you live in Europe, the answer is probably around 200 litres. That's enough to fill 16 big buckets. A typical north American uses around 400 litres a day for drinking, cooking, and washing. In poorer parts of the world, an average person uses just 10 litres of water a day.

## How safe?

Dirty water is a killer. Worldwide, around 4000 children die every day from diarrhoea caused by unclean water and poor sanitation.

Water companies supply water to most UK homes. All this water is safe to drink. It has low levels of microbes, including disease-causing bacteria. It also has low levels of dissolved salts, such as nitrates, which are harmful to health.

## Making water safe to drink

In the UK, water companies supply water of the correct quality by

• choosing appropriate water sources
• filtering the water to remove solids
• sterilising the water with chlorine or ozone.

### Filtering

Water companies use many types of filtration, including:

• **Microstrainers** – large rotating sieves which remove solids, including algae, from the water.
• **Sand filters** – football-pitch-sized beds of sand about a metre deep. As the water trickles through the sand, solids and harmful bacteria are removed from the water.

### Sterilising

Two gases are used to **sterilise** (kill bacteria) in water:

• Chlorine kills bacteria. The chlorine remains dissolved in the water until it comes out of the taps in our homes, so it also kills any bacteria that get into the water on its journey through the pipes from the water treatment works.
• Ozone, $O_3$, kills bacteria and destroys pesticides.

### Water sources

Water companies take water from a variety of sources. Water from different sources needs different treatments.

Water from this borehole has been filtered by underground rocks for many years. It occasionally contains bacteria, so is sterilised with chlorine.

▲ Water from streams fills this storage reservoir. The water is contaminated by tiny bits of soil, algae, viruses, and bacteria. It passes through microstrainers and sand filters. Then chlorine is added, to destroy bacteria and viruses.

▲ Water from this canal is of poor quality. It contains algae, viruses, bacteria, and foul-smelling ammonia, and is a brown colour. It is filtered to remove solids. Activated carbon removes pesticides and ammonia from the water. Ozone is bubbled through the water to kill bacteria and destroy pesticides. Chlorine is added to keep the water clean on its journey to the user.

## Questions

1  List the three stages by which UK water companies produce water of the correct quality.

2  Give the purpose of filtering water through a sand filter.

↓ E

3  Explain why water from a borehole often needs only little treatment to make it safe to drink.

4  Explain why chlorine can keep water safe to drink throughout its journey from the water treatment works to people's homes.

↓ C

5  Describe and explain all the stages in making poor quality canal water safe to drink.

↓ A*

### Did you know...?

The weight of water that women in Africa and Asia carry on their heads is commonly 20 kg, the same as a typical UK airline luggage allowance.

## Learning objectives

After studying this topic, you should be able to:

✔ evaluate the impacts of adding fluoride to water

✔ compare home water filters

## Key words

**fluoridated water, activated carbon, adsorption**

## Did you know...?

In one British city, one in five five-year-olds has had a tooth extracted, and more than half have some tooth decay.

## Fluoride fracas

In the USA and Australia, about 70% of homes are supplied with **fluoridated water** – water that has had sodium fluoride, or another fluoride compound, added to it. In the UK, the figure is closer to 10%. Only people living in Yorkshire, Tyneside, and parts of the West Midlands have fluoridated water. There are arguments for and against adding fluoride to water.

> I am a dentist. Fluoride water helps prevent tooth decay. It means fewer children have toothache. I think every UK home should have fluoridated water.

> I compared dental health in the Irish Republic and Northern Ireland. In the Irish Republic, 70% of water is fluoridated. The average number of decayed, missing, and filled teeth per child is 1.3. In Northern Ireland, there is no fluoridation. The average number of decayed, missing, and filled teeth per child is 2.3.

> I think that adding fluoride to water is an expensive way of improving the dental health of just a few people. If everyone cleaned their teeth properly, and didn't eat sweets, there would be no need for water fluoridation.

> I don't think water should be fluoridated. When I was a little girl, my teeth were discoloured. The dentist said I had consumed too much fluoride. I know this was because I used to swallow my toothpaste, but the fluoride in water surely didn't help.

> I am a health worker in a poor area of the UK. I think we should add fluoride to the water just in areas where a high percentage of people have decayed teeth. This would be less expensive than adding fluoride to all water, but would still improve dental health.

**A** Describe one benefit of adding sodium fluoride to water.

**B** Suggest an economic reason for not supplying fluoridated water to all UK homes.

## Home water filters

All UK mains water is safe to drink. But some people want to improve the taste or quality of their tap water. They do this by using special filters or ion exchange resins.

## Carbon filters

Carbon water filters contain **activated carbon**. Activated carbon is produced from materials such as coconut shells. It has been processed to give it lots of tiny holes. This means it has a very large surface area. One gram of activated carbon can have a surface area of up to 1500 m². As tap water passes through a carbon filter, molecules of unwanted substances stick to the surface of the carbon. The process is called **adsorption** and the molecules are described as being adsorbed. Carbon filters remove chlorine from water, as well as some compounds with bad tastes or smells.

## Silver filters

Silver has long been known to help make water safe to drink. More than two thousand years ago, Cyrus the Great of Persia kept his drinking water in silver containers. In the 1960s, NASA used silver to help produce drinking water aboard the Apollo spacecraft. Today, some types of home water filters include a source of silver ions, $Ag^+$. The ions destroy many types of bacteria.

▲ Activated carbon, as viewed under an electron microscope

▲ This home water filter contains silver nanoparticles to destroy dangerous bacteria

## Ion exchange resins

Some home water treatment systems include an ion exchange resin. These are similar to ion exchange resins used to remove water hardness (see spread C3.8). They replace undesirable metal ions such as lead, copper, or cadmium ions that may be in the water with hydrogen, sodium, or potassium ions.

## Questions

1 What do activated carbon filters remove from tap water?

2 Identify one benefit of a water filter that contains a source of silver ions.

3 Draw up a table to summarise the substances removed from water by activated carbon filters, silver filters, and ion exchange columns.

4 Write a paragraph to describe and evaluate the economic and social arguments for and against supplying all UK homes with fluoridated water.

## Learning objectives

After studying this topic, you should be able to:

✔ describe and evaluate the production of pure water by distillation

## Key words

desalination, distillation

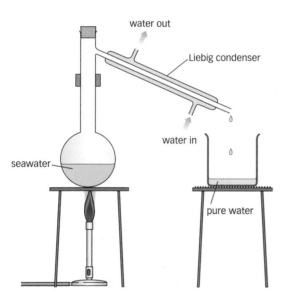

▲ You can use apparatus like this to produce pure water from seawater in the lab

**A** Write these stages of distillation in the order in which they happen: condense, evaporate, boil.

**B** What will remain in the round-bottomed flask at the end of the distillation process?

## Water shortage

The United Arab Emirates (UAE) has huge oil reserves. It is home to the world's tallest building, in Dubai. Many of its people have a lot of money to spend.

The demand for water in the UAE is high – average use in one region is 550 litres per person per day. But there are no permanent rivers or lakes in the United Arab Emirates, and rainfall is low in coastal areas. So where do people get their water from?

▲ Dubai at night

## Drinking seawater

Around 98% of water in the UAE comes from the sea. **Desalination** removes all the dissolved salts from the water, making it safe to drink.

There are several methods of desalination. The most important is **distillation**. In the laboratory, you can use the apparatus on the left to distil seawater.

Distillation happens on a huge scale in the largest desalination plant in the world, at Jebel Ali in the UAE. The process is similar to laboratory distillation, but at Jebel Ali the pressure of the system is reduced so that the water boils at a lower temperature. This reduces the energy costs of the process.

▲ The largest desalination plant in the world, at Jebel Ali

▲ Ships and submarines produce pure water by distillation, too

## Pros and cons

Distillation is a vital source of water for many, but it has its disadvantages:

- The water produced is completely pure. It contains no dissolved salts. So people drinking it do not experience the health benefits of dissolved calcium and magnesium ions.
- Many people dislike the taste of completely pure water.
- Distillation requires huge energy inputs, so the economic and environmental costs of distillation plants can be high.
- Sea life may be damaged at the water intake, and where highly concentrated salty water is returned to the sea.

### Questions

1 What is desalination?

2 Describe and explain the processes that happen in the distillation of seawater to produce pure water.

3 Draw up a table to summarise the environmental and economic disadvantages of producing drinking water from seawater.

4 Imagine that the government of a country on the coast of South America is considering building a desalination plant to supply its people with water. Write a paragraph to help the government weigh up the pros and cons of this proposal.

▼E

↓C

↓A*

### Exam tip | AQA

✔ Remember – distillation requires huge amounts of energy. This makes it very expensive as a means of supplying drinking water on a large scale.

## Learning objectives

After studying this topic, you should be able to:

✔ describe how to measure the energy transferred when foods and fuels burn

| TYPICAL NUTRITIONAL VALUES | | |
|---|---|---|
| | Per 34.5 g pack | Per 100 g |
| Energy | 761 kJ | 2207 kJ |
| | 163 kcal | 529 kcal |
| Protein | 2.0 g | 5.9 g |
| Carbohydrate | 17.1 g | 49.7 g |
| of which sugars | 0.1 g | 0.4 g |
| Fat | 11.8 g | 34.2 g |
| of which saturates | 0.9 g | 2.5 g |
| of which mono-unsaturates | 0.6 g | 27.9 g |
| of which polyunsaturates | 0.8 g | 2.2 g |
| Fibre | 1.4 g | 4.2 g |
| Sodium* | 0.21 g | 0.60 g |
| *Equivalent as salt | 0.53 g | 1.17 g |

**A** A **joule** is the unit of energy. Give the number of joules in one **kilojoule**, 1 kJ.

**B** Which stores more energy – 100 g of crisps or 100 g of cashews?

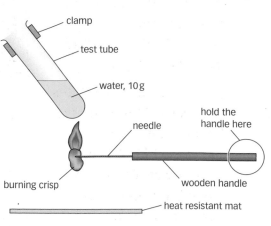

Measuring the energy in food

## Behind the label

Which do you prefer – crisps or cashew nuts? Which provide more energy?

| Per 30 g serving | | | | |
|---|---|---|---|---|
| Calories | Fat | Saturates | Salt | Sugars |
| 175 | 14.5 g | 2.7 g | 0.0 g | 1.7 g |

| Nutrition | | |
|---|---|---|
| Typical values | Per 30 g serving | Per 100 g |
| Energy | 725 kJ | 2410 kJ |
| | 175 kcal | 585 kcal |
| Protein | 5.7 g | 19.0 g |
| Carbohydrate | 5.3 g | 17.8 g |
| of which sugars | 1.7 g | 5.8 g |
| Fat | 14.5 g | 48.2 g |
| of which saturates | 2.7 g | 8.9 g |
| Fibre | 2.7 g | 8.9 g |
| Sodium | 0.0 g | trace |

Food labels tell us how much energy foods provide. Eating 100 g of cashew nuts provides 2410 kJ of energy, and 100 g of crisps provides 2207 kJ.

Of course, energy values are not the full story. Cashew nuts and crisps provide similar amounts of energy, but the nuts are more nutritious.

How do food companies know what numbers to write on the labels? Today, they use data tables to work out the energy values of processed foods. Before these data were available, scientists compared food energy values by using burning foods to heat water. The energy transferred on burning is similar to that available to the person eating the food.

## Measuring food energy

Freya pours 100 g of water into a metal container. She measures its temperature. She heats the water with a burning crisp. She measures the temperature again.

Here is a summary of Freya's results.

| Mass of crisp (g) | 1 |
|---|---|
| Increase in water temperature (°C) | 40 |

Freya uses an equation to calculate the heat energy, $Q$, transferred to the water:

$$Q = mc\Delta T$$

- $m$ is the mass of water, in grams.
- $c$ is the **specific heat capacity** of the water. It is the energy needed to make 1 g of water 1 °C hotter. Its value is 4.2 J/g°C.
- $\Delta T$ is the temperature change of the water, in °C.

So for Freya's experiment, the heat, $Q$, transferred to 100 g of water by 1 g of crisps:

$$= 100 \text{ g} \times 4.2 \text{ J/g°C} \times 40 °C$$
$$= 16\,800 \text{ J}$$
$$= 16.8 \text{ kJ}$$

This gives a value of –1680 kJ for burning 100 g of crisps. The negative sign shows that the burning reaction is exothermic. It transfers energy to the surroundings. In other words, it gives out energy.

Freya's value is different from that on the crisp packet. There are two reasons for this:

- Not all the heat from the burning crisp was transferred to the water – some was transferred to the surroundings and the apparatus.
- Some of the energy in crisps – that in the fibre – cannot be absorbed by the body. This energy is not included in the energy value on the crisp packet.

## Comparing fuels

You can use a similar experiment to compare the heat produced by burning fuels. The diagram shows how.

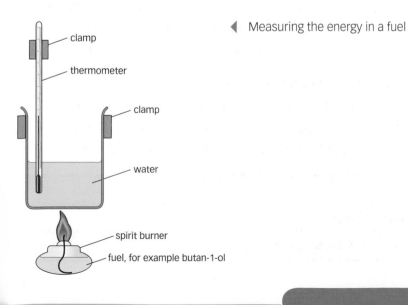

◀ Measuring the energy in a fuel

- clamp
- thermometer
- clamp
- water
- spirit burner
- fuel, for example butan-1-ol

## Key words

joule, kilojoule, specific heat capacity

## Did you know...?

100 g of crisps provides 10 times more energy than an apple of the same mass, but the apple is much richer in vitamins and minerals.

## Exam tip

✔ Take care with units, and note whether energy values are given in joules (J) or kilojoules (kJ). You may even be given energy data in calories.

## Questions

1 Give the symbol for the scientific unit of energy.

2 Eva burns 1 g of butan-1-ol fuel in the apparatus shown. It makes 100 g of water 55 °C hotter. Calculate the amount of heat energy transferred to the water from the butan-1-ol. The specific heat capacity of water is 4.2 J/g°C.

3 Eva checks her result in a data book. This states that burning 1 g of butan-1-ol releases 36 122 J of heat energy. Suggest why Eva's value is different from that in the data book.

## Learning objectives

After studying this topic, you should be able to:

✔ calculate energy transfers for reactions in solution

## Key words

**exothermic, endothermic**

## How hot?

Combustion reactions can transfer huge amounts of energy to the surroundings.

▲ Forest fires give out huge amounts of energy

Other types of reaction give out heat energy too. They are **exothermic**. Barney adds magnesium powder to dilute hydrochloric acid. After a few minutes, the beaker feels warmer. So Barney knows the reaction is exothermic. But how much energy has the reaction transferred to the surroundings? Barney does an experiment to find out. He

- pours 20 cm³ of hydrochloric acid into an insulated container
- measures the temperature of the acid
- adds the magnesium powder, with stirring
- observes the temperature change for a few minutes and records the maximum temperature reached.

Here are Barney's results:

| volume of acid (cm³) | 20 |
|---|---|
| temperature at start (°C) | 19 |
| highest temperature reached (°C) | 78 |
| temperature change (°C) | 59 |

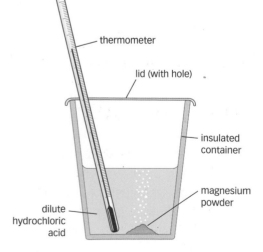

▲ Measuring energy transfer for the reaction of magnesium with dilute hydrochloric acid

Barney uses the equation $Q = mc\Delta T$ to calculate the energy transfer. He assumes that it is only the water in the solution that is being heated. So 4.2 J will raise the temperature of 1 g of the solution by 1 °C. He also assumes that the density of the solution is 1 g/cm³, so the mass of 20 cm³ of acid is 20 g.

$$Q = m \times c \times \Delta T$$
$$Q = 20\ g \times 4.2\ J/g\,°C \times 59\,°C$$
$$Q = 4956\ J$$

The energy change for the reaction is –4956 J for the amounts used in the experiment. The negative sign shows that the reaction is exothermic.

## How cold?

If you add citric acid powder to sodium hydrogencarbonate solution, the reacting mixtures cools down. Then, slowly, the mixture warms up until it reaches room temperature. The reaction has taken in heat energy from the surroundings. It is **endothermic**.

You can use the equation $Q = mc\Delta T$ to calculate how much energy the reaction mixture takes in from the surroundings. For endothermic reactions, the value of $Q$ is positive.

### Questions

1 Suggest why Barney uses an insulated container in his experiment.

2 Describe the difference between an exothermic reaction and an endothermic reaction.

⬇ E

3 Calculate the energy transferred by the reaction of aluminium powder with 20 cm³ of hydrochloric acid. The temperature increase was 76 °C.

4 Calculate the energy transferred by the reaction of citric acid with 20 cm³ of sodium hydrogencarbonate solution. The temperature decrease was 11 °C.

⬇ C

5 Suggest why the temperature of a reacting mixture in an endothermic reaction decreases at first, and then later returns to room temperature.

⬇ A*

A Calculate the energy transferred by the reaction of zinc powder with 20 cm³ of hydrochloric acid. The temperature increase was 9 °C.

B Calculate the energy transferred by the reaction of 100 cm³ of 1 mol/dm³ hydrochloric acid with 100 cm³ of 1 mol/dm³ sodium hydroxide solution. The temperature increase was 6 °C.

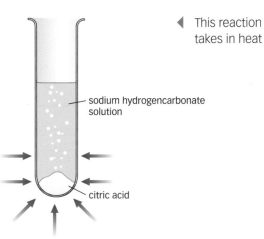

◀ This reaction takes in heat

sodium hydrogencarbonate solution

citric acid

### Did you know...?

Photosynthesis is endothermic. Plants absorb energy in the form of light. They can then convert carbon dioxide and water into glucose and oxygen.

### Exam tip

✔ To remember the difference between exothermic and endothermic reactions, think of *ent*ering – or going *in* to – a room. *End*othermic reactions take *in* energy.

## Key words

energy level diagram, activation energy

---

**A** Use the energy level diagram to work out the energy change for reaction 2.

**B** Use the two energy level diagrams to decide which of the two reactions transfers more energy to the surroundings.

## New hand warmer

Carmella is a chemist. She is developing a new type of hand warmer. She tries reacting different pairs of substances. Which would transfer most energy to a person's hands? The chemist draws **energy level diagrams** to represent her results. An energy level diagram shows the relative energies of the reactants and products in a reaction.

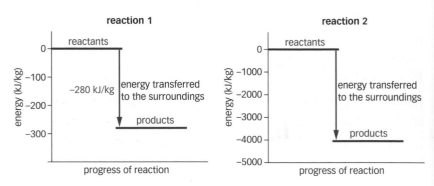

Energy level diagrams for two reactions. The two diagrams are drawn to different scales.

The energy level diagrams show that reactions 1 and 2 are exothermic. In both reactions, the energy stored in the products is less than the energy stored in the reactants. The extra energy first heats up the reaction mixture. Then it is transferred to the surroundings, as heat. Energy changes for exothermic reactions are negative. The energy change for reaction 1 is −280 kJ/kg.

## Sports injury pack?

The reaction represented by the energy level diagram on the left would be no good for a hand warmer. It would be better for a sports injury pack, to cool an injured arm or leg.

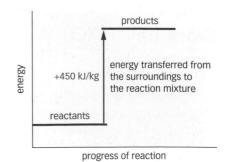

An energy level diagram for an endothermic reaction

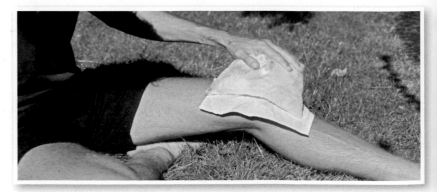

Sports injury packs cool injured limbs

The energy level diagram for the sports injury pack shows that the energy stored in the products is more than the energy stored in the reactants. The extra energy was taken from the reaction mixture. So its temperature fell. Then the mixture took in energy from the surroundings. Its temperature increased, back to room temperature. The reaction is endothermic. Energy changes for endothermic reactions have positive values.

The energy change for the reaction in the diagram is +450 kJ/kg.

# Getting going

All chemical reactions need energy to get them going. Reactions can only happen when reactant particles collide. Only those particles with enough energy are able to react when they do collide. The minimum energy needed for a reaction to start is the **activation energy**.

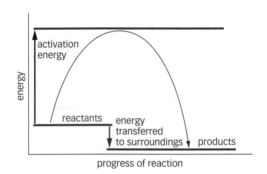

◀ Energy level diagram with activation energy. The curved arrow shows the energy as the reaction proceeds.

# Catalysts and activation energy

Catalysts speed up reactions without themselves being used up. Catalysts provide a different pathway for a reaction, with a lower activation energy.

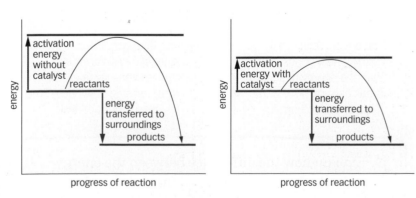

▲ The two energy level diagrams are for the same reaction. The diagram on the right shows that the activation energy is lower when a catalyst is used.

## Exam tip  AQA

✔ In energy level diagrams, the reactants are above the products for exothermic reactions. The products are above the reactants for endothermic reactions.

## Questions

1   What is a catalyst?

2   Explain the meaning of the term activation energy.

3   Sketch an energy level diagram that represents an exothermic reaction.

4   Draw an energy level diagram to represent the endothermic reaction in which nitrogen dioxide gas is formed from its elements. During the reaction, 34 kJ/mol is absorbed from the surroundings.

5   Use the diagram on the left to help you explain how catalysts reduce activation energy.

## Learning objectives

After studying this topic, you should be able to:

✔ explain energy changes in terms of making and breaking bonds

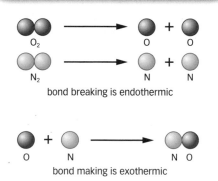

bond breaking is endothermic

bond making is exothermic

▲ Bond breaking in $N_2$ and $O_2$ and bond making in NO

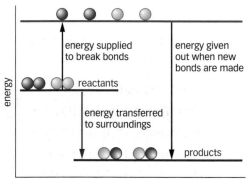

▲ Bond breaking is endothermic. Bond making is exothermic.

**A** Is bond breaking exothermic or endothermic?

## Nasty NO$_x$

Every year, UK cars must pass an exhaust emissions test. There are strict limits on amounts of exhaust pollutants, including oxides of nitrogen, NO$_x$. Oxides of nitrogen form when nitrogen and oxygen from the air react together. At normal temperatures, there is no reaction. But inside a hot car engine, the gases react rapidly.

## Bond breaking, bond making

A minimum energy, the activation energy, is needed to start the reaction. Heat from the car engine supplies this energy. It is used to break bonds in nitrogen molecules, $N_2$, and oxygen molecules, $O_2$. Bond breaking is endothermic.

New bonds are made as products such as nitrogen monoxide, NO, are formed. Bond making releases energy. It is an exothermic process.

## Exothermic or endothermic?

The difference between the energy supplied to break bonds in reactants and the energy released on forming bonds in products determines whether a reaction is exothermic or endothermic.

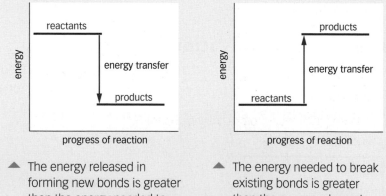

▲ The energy released in forming new bonds is greater than the energy needed to break existing bonds. The reaction is exothermic.

▲ The energy needed to break existing bonds is greater than the energy released from forming new bonds. The reaction is endothermic.

**B** Explain how the difference between the energy supplied to break bonds and the energy released on forming bonds determines whether a reaction is endothermic or exothermic.

Bonds are made as nitrogen oxides form in car exhaust

**Key words**

bond energy

**Exam tip** AQA

✔ Energy must be supplied to break bonds. Energy is released when bonds are formed.

## Using bond energies

Every type of bond needs a certain amount of energy in order to break. This is the **bond energy**. You can use bond energy data to calculate energy transfers in reactions.

| Bond | Bond energy (kJ/mol) |
|------|----------------------|
| H–H | 436 |
| Cl–Cl | 243 |
| F–F | 158 |
| H–Cl | 432 |
| H–F | 562 |

## Worked example

Calculate the energy change for the reaction

$$H_2(g) + Cl_2(g) \rightarrow 2HCl(g)$$

One mole of H–H bonds and one mole of Cl–Cl bonds break in the reaction.

The energy needed to break these bonds is
(436 + 243) = 679 kJ.

Two moles of H–Cl bonds form in the reaction. The energy released by this process is (2 × 432) = 864 kJ.

The overall energy transfer
= the energy supplied to break bonds – the energy released on making bonds
= 679 – 864
= –185 kJ

The negative sign shows that, overall, the reaction is exothermic.

## Questions

1 Is bond making endothermic or exothermic? ↓ E

2 An ozone molecule, $O_3$, splits up to make an oxygen molecule, $O_2$, and an oxygen atom, O. Predict whether the process is endothermic or exothermic. Give a reason for your answer. ↓ C

3 Use bond energy data to calculate the energy change for the reaction
$$H_2(g) + F_2(g) \rightarrow 2HF(g)$$

4 Which process is more exothermic – the formation of hydrogen chloride gas from its elements, or the formation of hydrogen fluoride gas from its elements? ↓ A*

# 16: Hydrogen fuel

## Learning objectives

After studying this topic, you should be able to:

✔ evaluate the use of hydrogen as a fuel for cars

## Key words

particulate, internal combustion engine, fuel cell

## Problems with petrol

Loretta loves her car. She says it makes travel quick and convenient. But her son, Seth, thinks his mum should cycle everywhere, like he does. Burning petrol and diesel in car engines produces damaging exhaust products:

- Carbon dioxide – causes climate change.
- **Particulates** (tiny particles of soot and unburned fuel) – may lead to asthma, lung cancer, and heart disease. They are created by burning diesel.
- Oxides of nitrogen – cause acid rain. They also destroy the ozone in the upper atmosphere that protects us from cancer-causing ultraviolet radiation.

## Alternatives to oil

We cannot fuel cars with petrol and diesel forever. Petrol and diesel are produced from crude oil. Supplies of crude oil are finite – they will one day run out.

One alternative to fossil fuel is hydrogen. Hydrogen cars produce mainly one exhaust product:

$$\text{hydrogen} + \text{oxygen} \rightarrow \text{water}$$
$$2H_2(g) + O_2(g) \rightarrow 2H_2O(g)$$

## Two types of hydrogen vehicle

Hydrogen-fuelled cars are in an early stage of development. There are two types:

- Some burn hydrogen in their **internal combustion engines**, such as Arnold Schwarzenegger's Hydrogen Hummer.
- Others, such as Honda's FC Sport, have hydrogen **fuel cells**. Hydrogen gas flows into the fuel cell. There, it reacts with oxygen. The process generates electricity to move the car.

A  List three disadvantages of fuelling cars with petrol and diesel.

B  Name the main exhaust product of hydrogen cars.

## Did you know...?

Arnold Schwarzenegger's Hydrogen Hummer needs refuelling every 60 miles.

▲ Schwarzenegger's Hydrogen Hummer

▲ The Honda FC Sport

# Sources of hydrogen

There is no naturally occurring hydrogen gas on Earth. So hydrogen fuel must be manufactured. Most hydrogen is made by reacting methane with water:

methane + water → hydrogen + carbon monoxide
$$CH_4(g) + H_2O(l) → 3H_2(g) + CO(g)$$

Carbon monoxide is poisonous, so it is reacted with oxygen as it is made. The product is carbon dioxide, a greenhouse gas.

carbon monoxide + oxygen → carbon dioxide
$$2CO(g) + O_2(g) → 2CO_2(g)$$

# Fuel cell versus internal combustion engine

Both types of hydrogen-powered vehicles have pros and cons.

| Fuel cells | Hydrogen-fuelled internal combustion engine (ICE) |
|---|---|
| more efficient than ICEs | less efficient than fuel cells |
| batteries expensive to produce, but getting cheaper as the process is automated | technology well understood, since most cars have them |
| include an expensive platinum catalyst, but new battery designs require less platinum | nitrogen and oxygen react in the engine to produce oxides of nitrogen as an exhaust gas, as well as water |
| few fuel stations currently supply hydrogen gas for refuelling | |
| methane, from which hydrogen is made, can be a renewable resource | |

## Questions

1 Name the two types of hydrogen car.

2 Give two advantages of hydrogen cars compared to petrol and diesel cars.

3 Describe how hydrogen is manufactured from methane gas.

4 Write a paragraph to compare the advantages and disadvantages of the combustion of hydrogen in car engines with the use of hydrogen fuel cells.

E

↓ C

↓ A*

## Exam tip AQA

✔ In the exam, you may be given information about the two types of hydrogen car, and asked to compare their advantages and disadvantages.

# Course catch-up

## Revision checklist

- Newlands listed elements in order of atomic weights. Mendeleev arranged elements into groups and periods to fit repeating patterns in properties.
- Modern periodic tables arrange elements by atomic number.
- Number of electrons in the highest energy level indicates group number.
- Elements in Group 1 (alkali metals) are soft, low-density metals. Alkali metals react rapidly with water, forming alkalis and hydrogen. Alkali metals form ionic compounds with non-metals.
- Group 1 ions have a charge of 1+.
- Group 1 elements become more reactive further down the group.
- Transition elements are denser, stronger, harder, and less reactive than alkali metals. They form coloured compounds and may act as catalysts. Transition metals have ions with different charges.
- Elements in Group 7 (halogens) are coloured non-metals.
- Reactive halogens displace less reactive halogens from solutions of halide ions.
- Group 7 ions have a charge of 1–.
- Group 7 elements become less reactive further down the group.
- Hard water is caused by calcium and magnesium ions dissolving when acidic rainwater flows through rocks.
- Temporary water hardness (caused by calcium hydrogencarbonate) can be removed by boiling. Permanent hardness (caused by calcium sulfate) isn't removed by boiling.
- Washing soda or ion-exchange resins also soften water.
- Water is filtered to remove solids and sterilised with chlorine to kill microbes. Dissolved substances are removed by specialised filters or ion exchange.
- Seawater can be desalinated by distillation, which needs a lot of energy.
- In calorimetry, the energy released from a chemical reaction is transferred to water. Energy transferred = $mc\Delta T$.
- Energy level diagrams show how the energies of chemicals change during a reaction. Energy is released when chemical bonds form and is required to break bonds.
- Catalysts reduce the minimum amount of energy needed to start a reaction (the activation energy).
- Hydrogen releases energy when it reacts with oxygen in combustion or in fuel cells.

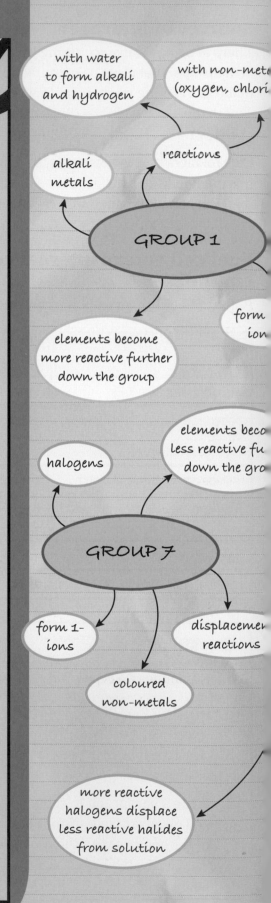

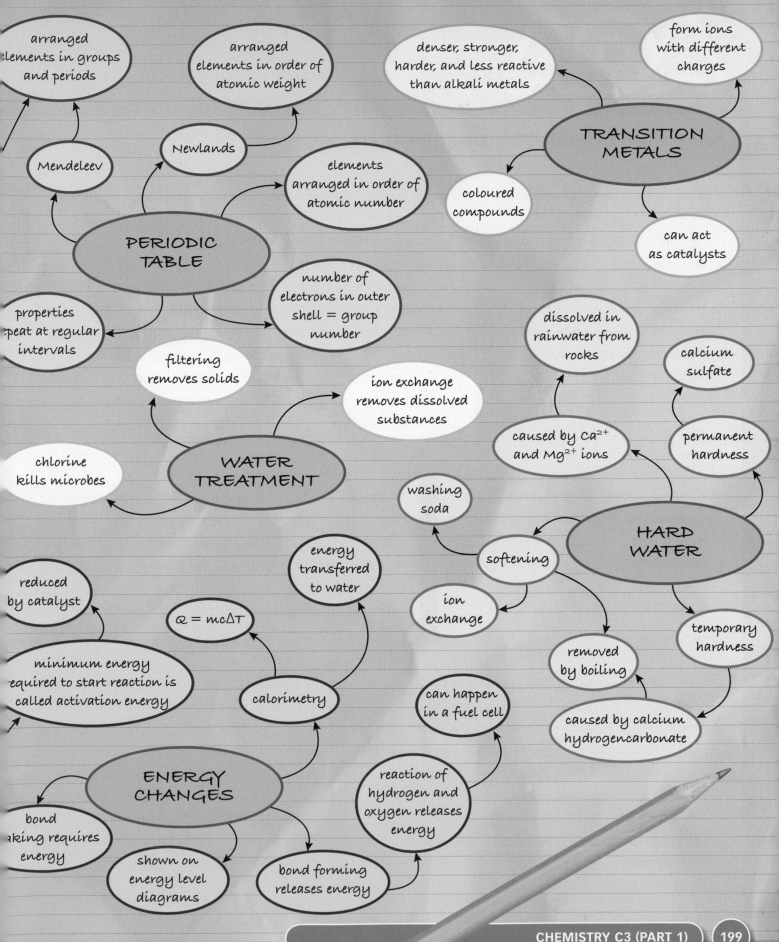

arranged
elements in groups
and periods

arranged
elements in order of
atomic weight

denser, stronger,
harder, and less reactive
than alkali metals

form ions
with different
charges

Newlands

Mendeleev

elements
arranged in order of
atomic number

TRANSITION
METALS

coloured
compounds

can act
as catalysts

PERIODIC
TABLE

number of
electrons in outer
shell = group
number

properties
repeat at regular
intervals

filtering
removes solids

ion exchange
removes dissolved
substances

dissolved in
rainwater from
rocks

calcium
sulfate

chlorine
kills microbes

WATER
TREATMENT

caused by $Ca^{2+}$
and $Mg^{2+}$ ions

permanent
hardness

washing
soda

softening

HARD
WATER

reduced
by catalyst

energy
transferred
to water

ion
exchange

temporary
hardness

$Q = mc\Delta T$

minimum energy
required to start reaction is
called activation energy

calorimetry

can happen
in a fuel cell

removed
by boiling

caused by calcium
hydrogencarbonate

ENERGY
CHANGES

reaction of
hydrogen and
oxygen releases
energy

bond
making requires
energy

shown on
energy level
diagrams

bond forming
releases energy

## Answering Extended Writing questions

Water in some parts of the UK is described as hard. What causes hard water, and what problems can hard water cause?

**The quality of written communication will be assessed in your answer to this question.**

---

**G–E**

Hard water means that there are rocks that have got into the water. You can't wash so easily in hard water so people don't like it.

**Examiner:** The candidate knows something about hard water, but hasn't explained it very clearly. It is important to mention that dissolved substances cause hard water. It is harder to wash in hard water, but the candidate should mention that this is a problem when soap is used. The candidate does not use any technical terms.

---

**D–C**

Water is hard because of calcium and magnesium dissolved in it which come from rocks. It causes scale and scum from soap which is a problem in people's homes. It can be removed by iron-exchange or washing soda.

**Examiner:** The candidate deals with both aspects of the question but does not provide much detail about either – scale and scum are not explained, for example, and it is calcium and magnesium ions that cause hard water. The sentence about treatment is not relevant to the question. One spelling error.

---

**B–A\***

Hard water is caused by the presence of dissolved calcium and magnesium compounds. These can get into water because when acidic rain water passes through rocks, substances like calcium hydrogencarbonate will dissolve into the water. Hard water causes problems because it can make scale when it is heated which clogs up heating systems and kettles. It also reacts with soap to make scum.

**Examiner:** This answer includes an excellent range of facts that cover both aspects of the question. The section about the problems caused could be expanded to explain that scale makes heating systems inefficient and scum makes soap less effective. Both of these increase costs to the consumer. Spelling, punctuation, and grammar are good.

# Exam-style questions

**1 a** Complete the following description of the periodic table, using the words below.

masses    numbers    groups
periods    properties

The periodic table contains elements arranged in order of their atomic _____. It contains vertical columns called _____ and horizontal rows called _____. Elements in the same group have similar _____.

**b** Name an element from:
  **i** Group 1    **ii** Group 7
  **iii** the transition elements.

**2** Kate adds 1 g of zinc to 100 cm³ of dilute sulfuric acid in an insulated cup. She measures a temperature rise of 23.6 °C.

**a** Use the equation $Q = mc\Delta T$ to calculate the amount of energy in J released in this experiment ($c = 4.2$ J/g°C).

**b i** Copy and complete this energy level diagram to show the relative energies of the reactants and products.

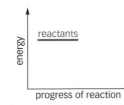

**ii** Label the diagram to show the energy change and the activation energy of the reaction.

**c** The reaction is faster with a copper sulfate catalyst. Explain why.

**3** Elements in Group 7 have similar properties. However, their reactivity increases going down the group.

**a** Halogen elements react by gaining electrons to form a halide ion.
  **i** Complete this equation to show how chlorine forms chloride ions: $Cl_2 + \_\_e- \rightarrow \_\_Cl-$
  **ii** What name is given to reactions in which electrons are gained?
  **iii** Bromine is less reactive than chlorine. Use ideas about electronic structure to explain why.

**b** When chlorine is added to a solution of potassium bromide, an orange solution is formed.
  **i** Name the orange product.
  **ii** What type of reaction has occurred?
  **iii** Complete and balance this symbol equation for the reaction: $Cl_2 + 2KBr \rightarrow \_\_$

## Extended Writing

**4** Water must be treated before it is suitable for drinking. Describe the treatments used in the UK water supply.

**5** Stefan lives in a part of the country with hard water. Describe how the water becomes hard and discuss the ways in which hard water may be softened.

**6** The work of Dmitri Mendeleev was important in developing the modern periodic table. Describe his contribution.

A01  Recall the science
A02  Apply your knowledge
A03  Evaluate and analyse the evidence

CHEMISTRY C3 (PART 1)    201

# C3 Part 2

# Analysis, ammonia, and organic chemistry

## Why study this unit?

Every year, UK factories make more than one million tonnes of ammonia. The gas makes vital fertilisers, explosives, and cleaning materials. We also use large amounts of carbon-based organic compounds, such as alcohols, carboxylic acids, and esters. Their properties make them ideal for many purposes. Health and environment workers and forensic scientists use chemistry to identify substances.

In this unit you will discover how to identify positive and negative ions in salts, and how to measure the amounts of substances in solution. You will also learn about the manufacture of ammonia, and how chemical engineers choose optimal conditions for the process. Finally, you will study patterns in the properties of three groups of organic compounds, and discover how their uses depend on their properties.

## You should remember

1  A salt is a compound that contains metal ions, and that can be made from an acid.

2  Concentration is the amount of a substance in a certain volume of solution.

3  A reversible reaction is one in which the products of a reaction can react to produce the original reactants.

4  Compounds of carbon are called organic compounds.

5  Organic compounds are classified into groups with similar properties, such as alkanes and alkenes.

This is a coloured magnetic resonance imaging (MRI) scan of a brain with alcoholic dementia. The condition is caused by drinking too much alcohol over many years. The brain has shrunk, so there is gap between the brain and the skull around the outside. Alcoholic dementia causes memory loss, confusion, and personality changes.

▲ Firework colours come from burning metal compounds

**A** Draw a table showing the flame colours given by lithium, sodium, potassium, calcium, and barium compounds.

▲ Iron(II) ions ($Fe^{2+}$) form a green precipitate. Iron(III) ions ($Fe^{3+}$) form a brown precipitate. Copper(II) ions ($Cu^{2+}$) form a blue precipitate.

## Firework fantasy

It's 5th November. Fireworks light up the sky. But how do chemists give fireworks their colours?

Firework colours come from burning metal compounds. Different metal ions give different coloured flames.

## Flame tests

Flame colours are not just useful for fireworks. They also help to identify metal ions. Several metal ions produce distinctive colours in **flame tests**. You can do a flame test by dipping the end of clean nichrome wire in a sample of a salt. Hold the end of the wire in a Bunsen flame, and observe the flame colour.

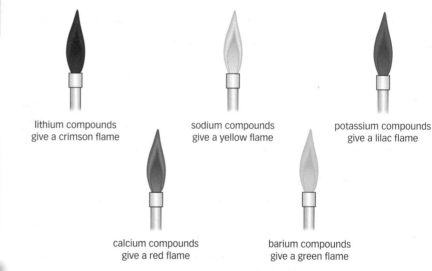

lithium compounds give a crimson flame

sodium compounds give a yellow flame

potassium compounds give a lilac flame

calcium compounds give a red flame

barium compounds give a green flame

**B** Apart from identifying metal ions, how else are flame colours useful?

## Identifying other metal ions

You can use sodium hydroxide to identify some metal ions. Here's how:

• Dissolve a sample of the compound in pure water to make a solution.
• Add a few drops of dilute sodium hydroxide to the solution.
• If a precipitate forms, observe its colour.

Different metal ions form precipitates of different colours.

The precipitates are all metal hydroxides. The equations show how copper(II) hydroxide is formed in a precipitation reaction:

| copper(II) chloride | + | sodium hydroxide | → | copper(II) hydroxide | + | sodium chloride |
|---|---|---|---|---|---|---|

$$CuCl_2(aq) + 2NaOH(aq) \rightarrow Cu(OH)_2(s) + 2NaCl(aq)$$

An ionic equation summarises the reaction. It shows only the ions involved in the reaction:

$$Cu^{2+}(aq) + 2OH^-(aq) \rightarrow Cu(OH)_2(s)$$

Aluminium hydroxide, calcium hydroxide, and magnesium hydroxide are all white precipitates. You can distinguish aluminium from the other two hydroxides by adding extra sodium hydroxide solution. Aluminium hydroxide dissolves in the excess sodium hydroxide:

$$Al(OH)_3(s) + OH^-(aq) \rightarrow Al(OH)_4^-(aq)$$

Precipitates of calcium hydroxide and magnesium hydroxide do not dissolve in excess sodium hydroxide.

▲ Aluminium ions ($Al^{3+}$), calcium ions ($Ca^{2+}$), and magnesium ions ($Mg^{2+}$) form white precipitates with sodium hydroxide solution

## Questions

1 Name the flame colours given by burning calcium and barium compounds.

2 Draw a table to show the results of adding dilute sodium hydroxide to solutions containing these ions:

$Cu^{2+}$   $Fe^{2+}$   $Fe^{3+}$   $Mg^{2+}$   $Ca^{2+}$   $Al^{3+}$

3 Write an ionic equation to summarise the reaction of sodium hydroxide solution with iron(III) chloride solution.

4 You have a white solid. You dissolve some of it in pure water to make a colourless solution. Adding dilute sodium hydroxide gives a white precipitate. Describe how to find out whether the original solid is magnesium chloride, calcium chloride, or aluminium chloride. Give the results you would expect for each salt.

### Key words

flame test

### Did you know...?

Bunsen and Kirchhoff discovered the elements caesium and rubidium by examining flame colours carefully.

### Exam tip

✔ In the exam, you may be asked to interpret flame test results and sodium hydroxide test results.

# 18: Identifying negative ions

## Did you know...?

In 1949, an American woman murdered her neighbour, and rival in love, by adding sodium fluoride-based insecticide to her coffee.

## Fluoride fatalities

Have you ever put salt in your tea, or sugar on your chips? These mistakes don't taste good, but are unlikely to harm you. However, some kitchen mix-ups can be fatal. In 1942, an American hospital cook added sodium fluoride – a cockroach killer – to scrambled eggs, instead of milk powder. Hours later, 47 patients were dead.

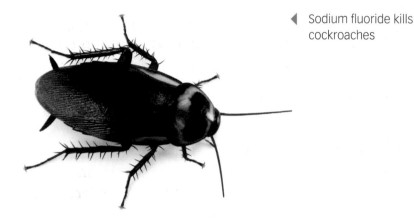

◀ Sodium fluoride kills cockroaches

You cannot identify a compound by looking at it. So chemists have devised tests and instrumental methods to identify chemicals. The previous spread describes tests for metal ions. This spread describes tests for negative ions.

## Testing for carbonates

To find out if a solid contains carbonate ions ($CO_3^{2-}$):

* Add a few drops of dilute hydrochloric acid to the surface of the solid.
* Watch carefully. If you notice fizzing, a gas is being produced.
* Use limewater to test the gas. If the limewater goes cloudy, the gas is carbon dioxide, and the solid is a carbonate.

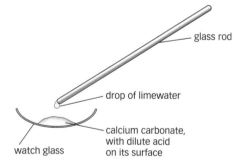

glass rod

drop of limewater

calcium carbonate, with dilute acid on its surface

watch glass

▲ Carbonates react with dilute acids to form carbon dioxide gas

For example:

calcium + hydrochloric → calcium + carbon + water
carbonate      acid         chloride    dioxide

$$CaCO_3(s) + 2HCl(aq) \rightarrow CaCl_2(aq) + CO_2(g) + H_2O(l)$$

---

**A** Name the products of the reaction of magnesium carbonate with hydrochloric acid.

**B** Describe how to test for carbon dioxide gas.

---

# Testing for halide ions

You can use silver nitrate solution to test for compounds containing chloride, bromide, and iodide ions. Here's how:

- Dissolve a sample of the solid in dilute nitric acid.
- Add silver nitrate solution.

Different halide ions form precipitates of different colours. The precipitates are silver halides. The equations below summarise the reaction of sodium chloride with silver nitrate:

| sodium chloride | + | silver nitrate | → | silver chloride | + | sodium nitrate |
|---|---|---|---|---|---|---|
| $NaCl(aq)$ | + | $AgNO_3(aq)$ | → | $AgCl(s)$ | + | $NaNO_3(aq)$ |

This ionic equation summarises the reaction:
$$Ag^+(aq) + Cl^-(aq) → AgCl(s)$$

# Testing for sulfates

To find out whether a compound includes sulfate ions ($SO_4^{2-}$):

- Dissolve a sample of the solid in dilute hydrochloric acid.
- Add barium chloride solution.

Sulfate compounds react with barium chloride solution to form a white precipitate of barium sulfate. The ionic equation for the reaction is:
$$Ba^{2+}(aq) + SO_4^{2-}(aq) → BaSO_4(s)$$

▲ Silver chloride (AgCl) is white. Silver bromide (AgBr) is cream. Silver iodide (AgI) is yellow.

## Exam tip

✔ Remember the precipitate colours – they will help you interpret the results of chemical analysis tests in the exam.

## Questions

1 Give the colour of the precipitate formed by reacting sodium chloride with acidified silver nitrate solution. ↓ E

2 Draw a table to show how to test for these ions: $CO_3^{2-}$, $SO_4^{2-}$, $Cl^-$, $Br^-$, $I^-$. In your table, include the names of the chemicals you add for each test, and the expected results. ↓ C

3 Write a balanced symbol equation for the reaction of magnesium sulfate with barium chloride in the presence of dilute hydrochloric acid.

4 Write an ionic equation for the reaction of sodium bromide with silver nitrate in the presence of dilute nitric acid. ↓ A*

## Learning objectives

After studying this topic, you should be able to:

- ✔ use the results of chemical tests to identify ions in salts
- ✔ describe how people in different jobs use chemical tests

▲ Environment workers collect and test water samples from rivers

## Health, the environment, and forensics

Paula is pregnant. She gives her midwife a urine sample. The midwife places a dipstick in the urine to test for protein. The dipstick changes colour if protein is present. Protein in the urine of a pregnant woman shows that she may have a serious condition that needs urgent treatment.

◀ Urine dipstick

Sophie works for an environment organisation. She collects river water samples. She takes the samples to a laboratory. There, scientists use instrumental methods to test the water for various substances, including lead ions. Lead ions may affect brain development in children. Lead ions in river water damage plants and animals, including fish.

Phil is a forensic scientist. He analyses drugs, such as heroin and cocaine, seized by the police. He uses instrumental methods to find out whether the drugs are pure, or mixed with other substances.

### Did you know...?

Forensic scientists test the blood and urine of people who suspect that their drink has been spiked with a mind-altering substance.

A Name three types of work that may involve identifying substances.

B Explain why it is important to measure the concentration of lead ions entering river water from a sewage treatment works.

# Identifying the substances in a mixture

Clare has a sample of small white crystals. She knows the sample is a mixture of two substances. She divides her sample in half. She dissolves one half in water, and leaves the other half as it is. Clare uses chemical tests to identify the ions in the mixture. The tables show the tests she does, and their results.

## Tests on the solid

| Description of test | Observations |
|---|---|
| flame test | bright yellow flame |
| add drops of dilute hydrochloric acid to the solid | no bubbles |

## Tests on the solution

| Description of test | Observations |
|---|---|
| add dilute nitric acid and silver nitrate solution | white precipitate |
| add dilute hydrochloric acid and barium chloride solution | white precipitate |
| add sodium hydroxide solution | white precipitate that dissolves in excess sodium hydroxide solution |

▲ Clare's sample produced a yellow flame

## Exam tip · AQA

✓ Health workers, environmental scientists, and forensic scientists use chemical tests or instrumental methods to identify chemicals.

◀ A solution of Clare's sample produced a white precipitate when dilute hydrochloric acid and barium chloride were added to it

## Questions

Use the two results tables above to help you answer the questions.

1   What does the flame test result show?

2   Explain how the test results show that the mixture includes no carbonate ions.

3   What do the results of the silver nitrate and barium chloride tests show?

4   Give the formulae of the four ions present in Clare's mixture.

5   Use ionic equations to help you explain the results of the silver nitrate, barium chloride, and sodium hydroxide tests.

E

C

A*

## Learning objectives

After studying this topic, you should be able to:

✔ use titrations to find out the volumes of solutions that react together

## Key words

**burette, end point, pipette, rough titration, titration**

▲ Is it safe for these boys to swim?

## Acid spill

It's 7.25 a.m. A park worker notices a pungent smell near an open-air swimming lake. He struggles to breathe. Fire crews arrive at the scene. They discover an acid spill. More than 150 litres of hydrochloric acid has leaked from the chlorination system that treats the lake water. The fire fighters close the swimming lake.

Later, public health officials turn up. They measure the concentration of acid in the lake water. It is safe for swimming. There is no need to add a base to neutralise extra acid in the water. Most of the acid must have soaked into the soil around the chlorination equipment.

## Measuring acids and alkalis

Environment workers sometimes add bases to lakes that have been acidified by acid rain. Beforehand, they use instrumental techniques to measure the concentration of acid in the lake water. They do calculations to work out how much base to add.

In the laboratory, you can do **titrations** to measure the volumes of acid and alkali solutions that react with each other. The diagrams show how to measure the volume of dilute hydrochloric acid that reacts with a 25.00 cm³ sample of dilute sodium hydroxide solution.

1. Use a pipette to measure accurately 25.00 cm³ of sodium hydroxide solution.

2. Allow the sodium hydroxide solution to run into a conical flask. Add a few drops of phenopthalein.

3. Pour hydrochloric acid into a burette. Read the scale. Add hydrochloric acid from the burette to the conical flask until the indicator just turns colourless. Read the scale. Calculate the amount of acid added. This rough titration gives an idea of how much acid is needed to neutralise the sodium hydroxide, or reach the end point.

4. Repeat steps 1–3, burette the acid one drop at a time as you near the end point. Swirl after each addition. Repeat until you have three consistent values for the acid volume.

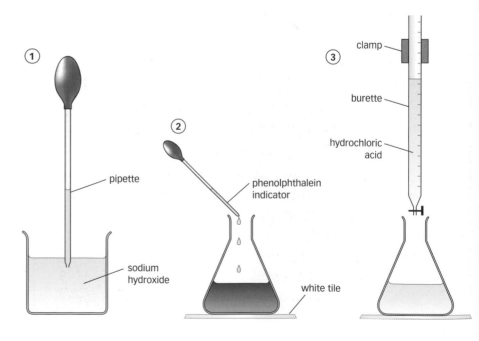

**A** What is the purpose of a rough titration?

**B** In a titration, what is the meaning of the term end point?

**Did you know...?**

Phenolphthalein is not the only indicator suitable for acid–base titrations. Depending on the acid and base, you can also use bromophenol blue, methyl orange, methyl red, bromothymol blue, and litmus. They all have different colour changes.

## Calculating titration volumes

The table gives titration results for neutralising 25.00 cm³ of 1.0 mol/dm³ sodium hydroxide solution with approximately 1 mol/dm³ hydrochloric acid.

|  | Rough | Run 1 | Run 2 | Run 3 |
|---|---|---|---|---|
| initial burette reading (cm³) | 0.25 | 25.15 | 0.05 | 24.25 |
| final burette reading (cm³) | 25.15 | 49.25 | 24.25 | 48.40 |
| volume of acid added (cm³) | 24.90 | 24.10 | 24.20 | 24.15 |

You can use the results to calculate the mean volume of acid added. The result of the rough titration is not included in the calculation.

Mean volume = (24.10 + 24.20 + 24.15) ÷ 3 = 24.15 cm³.

### Questions

1 Describe one purpose of doing titrations.

2 Name the piece of titration apparatus that has a tap and a graduated scale.

3 Rafat adds dilute sulfuric acid to a mixture of sodium hydroxide solution and litmus. What colour change will he see?

4 In a titration, if the initial burette reading is 0.75 cm³, and the final burette reading is 26.10 cm³, what volume of solution has been added?

5 Use the data below to calculate the mean volume of solution added from a burette.
Rough volume = 15.70 cm³
Run 1 volume = 15.30 cm³
Run 2 volume = 15.25 cm³
Run 3 volume = 15.30 cm³.

↓ E

↓ C

↓ A*

**Exam tip**

✔ Do not include rough titration results when calculating mean volumes in titrations.

## Learning objectives

After studying this topic, you should be able to:

✔ use titration results to calculate the concentrations of solutions

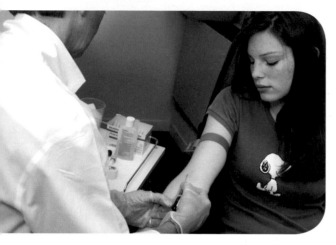

▲ Blood tests help health workers diagnose disease

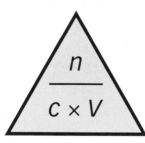

$$\frac{n}{c \times V}$$

▲ $n$ is the number of moles, or the mass, of the solute

A Calculate the concentration, in mol/dm³, of a solution that has 0.5 mol of an alkali in 2 dm³ of solution.

B Calculate the concentration, in mol/dm³, of a solution that has 2 mol of an acid in 250 cm³ of solution.

## Amounts matter

Lilya is ill. She has sickness and diarrhoea, and sweats at night. Her doctor orders blood tests. The test results show that the **concentration** of sodium ions in Lilya's blood is higher than normal. The concentration of potassium ions is lower than normal. The doctor uses these results to help him work out what is wrong with Lilya.

## Calculating concentration

The concentration of a solution is the amount of solute per unit volume of solution. A **solute** is a substance that is dissolved in solvent. Chemists measure concentration in grams per cubic decimetre (g/dm³), or in moles per cubic decimetre (mol/dm³). One dm³ is the same as one litre, or 1000 cm³. One mole (1 mol) of a substance is its formula mass in grams.

Use this equation to calculate concentration in g/dm³:

$$\text{concentration} = \frac{\text{mass of solute (in g)}}{\text{volume of solution (in dm}^3)}$$

Use this equation to calculate concentration in mol/dm³:

$$\text{concentration} = \frac{\text{number of moles of solute (in mol)}}{\text{volume of solution (in dm}^3)}$$

### Worked example 1

What is the concentration of a solution that has 36.5 g of solute in 500 cm³ of solution?

$$\text{volume of the solution in dm}^3 = \frac{500 \text{ cm}^3}{1000 \text{ cm}^3}$$

$$= 0.5 \text{ dm}^3$$

$$\text{concentration in g/dm}^3 = \frac{\text{mass}}{\text{volume}}$$

$$= \frac{36.5 \text{ g}}{0.5 \text{ dm}^3}$$

$$= 73 \text{ g/dm}^3$$

## Calculating masses and moles

If you know the concentration of a solution and its volume, you can calculate the number of moles of solute, or the mass of solute, in a given volume of solution.

▲ Titration apparatus

### Exam tip

✔ Take care with units – concentrations can be given in $g/dm^3$ or $mol/dm^3$.

### Questions

1 What is the concentration of a solution that has 40 g of solute in 2 $dm^3$ of solution?

2 What is the concentration of a solution that has 0.25 mol of solute in 125 $cm^3$ of solution?

3 How many moles of copper sulfate are there in 25 $cm^3$ of a 0.1 $mol/dm^3$ solution?

4 What mass of magnesium chloride is there in 1 $dm^3$ of a 1 $mol/dm^3$ solution? The formula of magnesium chloride is $MgCl_2$.

5 What mass of sodium fluoride is in 500 $cm^3$ of a 2 $mol/dm^3$ solution? The formula of sodium fluoride is NaF.

↓ A*

### Worked example 2

How many moles of sodium sulfate are there in 250 $cm^3$ of a 2 $mol/dm^3$ solution?

volume of the solution (in $dm^3$) = $\frac{250 \text{ cm}^3}{1000 \text{ cm}^3}$

$= 0.25 \text{ dm}^3$

number of moles = concentration in $mol/dm^3$ × volume in $dm^3$

$= 2 \text{ mol/dm}^3 × 0.25 \text{ dm}^3$

$= 0.5 \text{ mol}$

## Learning objectives

After studying this topic, you should be able to:

✔ use titrations to find the concentrations of acids and alkalis

---

**A** Describe two uses of titration reactions.

**B** Name two pieces of titration apparatus that measure volume accurately.

▲ Abdul doing titration

## Using titrations to calculate concentrations

Titrations are useful for measuring the volumes of acid and alkali solutions that react with each other. But this is not the only use of titrations. In titrations, if you know the concentration of one reactant, you can use titration results to find the concentration of the other reactant.

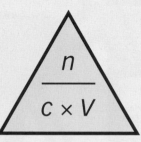

▲ $n$ is the number of moles, or the mass, of the solute

### Worked example

Abdul places 25.00 cm³ of sodium hydroxide solution in a flask. Its concentration is 0.50 mol/dm³. Abdul does a titration. He finds that it takes 12.70 cm³ of sulfuric acid to neutralise the sodium hydroxide solution. Calculate the concentration of the sulfuric acid in g/dm³.

Calculate the number of moles of sodium hydroxide:
number of moles
= concentration in mol/dm³ × volume in dm³
= 0.5 mol/dm³ × (12.7 ÷ 1000) dm³
= 0.00635 mol

Write a balanced equation for the reaction. Use it to work out the number of moles of sulfuric acid in 12.7 cm³ of solution.

$$2NaOH(aq) + H_2SO_4(aq) \rightarrow Na_2SO_4(aq) + 2H_2O(l)$$

The equation shows that 2 mol of sodium hydroxide reacts with 1 mol of sulfuric acid. Abdul has 0.00635 mol of sodium hydroxide.

So the number of moles of sulfuric acid in 12.7 cm³ is
(0.00635 ÷ 2) = 0.00318 mol of sulfuric acid.

Calculate the concentration of the sulfuric acid in mol/dm³

$$\text{concentration} = \frac{\text{number of moles}}{\text{volume}}$$

$$= \frac{0.00318 \text{ mol}}{(12.7 \div 1000) \text{ dm}^3}$$

$$= 0.25 \text{ mol/dm}^3$$

Calculate the concentration of the sulfuric acid in g/dm³

Mass of 1 mol of sulfuric acid,
$H_2SO_4 = (2 \times 1) + 32 + (4 \times 16)$
   $= 98$ g

So the mass of 0.25 mol of sulfuric acid
$= 0.25 \times 98$ g
$= 24.5$ g

So the concentration in g/dm³ = 24.5 g/dm³

### Did you know...?

The words pipette and burette were first used by the French chemist Joseph Gay-Lussac in 1824.

### Exam tip

✔ To convert moles to mass, multiply the number of moles by the formula mass.

## Questions

1 Harry places 25.00 cm³ of sodium hydroxide solution in a flask. Its concentration is 1.0 mol/dm³. It takes 27.00 cm³ of hydrochloric acid to neutralise the solution. Calculate the concentration of the hydrochloric acid, HCl, in g/cm³.

2 Sabina places 25.00 cm³ of nitric acid in a flask. Its concentration is 0.50 mol/dm³. It takes 28.0 cm³ of potassium hydroxide solution to neutralise the acid. Calculate the concentration of the potassium hydroxide in mol/dm³.

↓ A*

3 Marcus places 10.00 cm³ of sulfuric acid in a flask. Its concentration is 0.20 mol/dm³. It takes 19.00 cm³ of sodium hydroxide solution to neutralise the acid. Calculate the concentration of the sodium hydroxide in g/dm³.

## Learning objectives

After studying this topic, you should be able to:

✔ describe how ammonia is made

▲ Fertilisers made from ammonia hugely increase crop yields

▲ Ammonia gas is used in the refrigeration of some ice skating rinks

## Amazing ammonia

**Ammonia** is a toxic gas with a penetrating, suffocating smell. It may cause serious burns, and is extremely harmful to the eyes. Its reactions can be violent. Even so, UK factories make more than 1 million tonnes of the gas each year. Worldwide, factories in China, India, Russia, and other countries produce around 3 million tonnes of ammonia every week. Why?

▲ Ammonia factory

Ammonia is a compound of nitrogen and hydrogen. Its formula is $NH_3$. Plants need nitrogen to grow properly, so nitrogen compounds are important fertilisers. More than 80% of the world's ammonia is used to make fertiliser compounds, for example ammonium nitrate, $NH_4NO_3$.

Chemical companies also convert huge amounts of ammonia into nitric acid. The acid is used to make more fertilisers, and explosives such as TNT (trinitrotoluene). Dilute ammonia solution reacts with grease, so it makes good glass and oven cleaners.

## Raw materials

Ammonia is made from its elements, nitrogen and hydrogen:

- Nitrogen is separated from the air.
- There are several possible sources of hydrogen. One of these is natural gas. Methane from natural gas reacts with steam:

methane  +  steam  →  carbon monoxide  +  hydrogen
$$CH_4(aq)  +  H_2O(g)  →  CO(g)  +  3H_2(g)$$

**A** Give the formula of ammonia.

**B** Name the raw materials for making ammonia. Describe how they are obtained.

**Did you know...?**

Fritz Haber discovered how to make ammonia from nitrogen gas in 1909. He and his colleagues did more than 6500 experiments to find the best catalyst for the process.

## Making ammonia

Ammonia is made from its raw materials by the **Haber process**. The diagram on the right summarises the processes and conditions involved.

In the reaction vessel, hydrogen and nitrogen begin reacting to form ammonia. This is a **reversible reaction**. This means that, as some ammonia molecules are being made, others are breaking down to make hydrogen and nitrogen again. The ⇌ sign shows that the reaction is reversible:

$$\text{nitrogen} + \text{hydrogen} \rightleftharpoons \text{ammonia}$$
$$N_2(g) + 3H_2(g) \rightleftharpoons 2NH_3(g)$$

There are three gases in the reaction vessel. They are cooled and separated in the condenser. The table shows their boiling points.

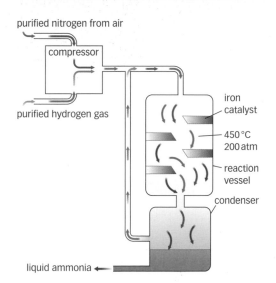

| Substance | Boiling point (°C) |
|-----------|--------------------|
| nitrogen  | −196 |
| hydrogen  | −252 |
| ammonia   | −33 |

▲ This diagram summarises the Haber process

## Questions

1 Describe two properties of ammonia.

2 Describe three uses of ammonia.

3 List the conditions for the Haber process. Include the temperature, pressure, and catalyst.

4 Explain the meaning of the term reversible reaction.

5 Study the boiling point data in the table. Use it to suggest how ammonia is separated from hydrogen and nitrogen in the condenser.

**Exam tip**   **AQA**

✔ Revise the Haber process carefully. It comes up in exams very often.

**Key words**

**ammonia, Haber process, reversible reaction**

## Money in ammonia

Making ammonia is big business. There is a huge demand for the gas. But the process is not cheap. To maximise profits, companies need to keep costs low. They also need to make as much product as possible, as quickly as possible.

▲ Most ammonia is used to make fertilisers, but some is used for making explosives

## Choosing conditions

As we saw on the previous spread, the Haber process involves a reversible reaction:

$$N_2(g) + 3H_2(g) \rightleftharpoons 2NH_3(g)$$

Two reactions are happening at once:

- a forward reaction, in which nitrogen and hydrogen react to form ammonia
- a backward reaction, in which ammonia breaks down to form hydrogen and nitrogen.

The reaction vessel always contains a mixture of nitrogen, hydrogen, and ammonia. Companies want conditions that maximise the amount of ammonia made in the reaction, or its **yield**. They also want ammonia to form quickly.

**A** Suggest why chemical companies want to maximise yield.

**B** Explain why operating at high pressure is expensive.

### Pressure

The higher the pressure, the higher the yield of ammonia. But the higher the pressure, the stronger the reaction vessel and pipes need to be. Strong vessels and pipes are expensive. The chemical engineers who design new Haber process plants compromise with a pressure of about 200 atmospheres.

## Temperature

The lower the temperature, the higher the yield of ammonia. But the reaction is slow at low temperatures. So there will be only small amounts of ammonia to sell each day. The chosen temperature, 450°C, is another compromise – this time between yield and speed.

## Catalyst

Catalysts increase reaction speeds without being used up in the reaction. Using a catalyst increases the yield at a given temperature and pressure. Iron is the chosen catalyst in the Haber process. It works well, and is cheap and easy to obtain. It is important to look after the catalyst. Tiny amounts of impurities can stop it working properly. That's why the hydrogen and nitrogen entering the reaction vessel must be very pure.

▲ Magnetite, a form of iron oxide, can also be used as a catalyst in the Haber process

## Energy costs

Keeping the reaction vessel at 450°C requires energy inputs. Once the process has started, much of the heat is supplied within the process. As the gas mixture cools in the condenser, energy is transferred to a **heat exchanger**. The heat exchanger heats up the nitrogen and hydrogen before they enter the reaction vessel.

Minimising energy inputs helps to minimise the impacts of ammonia production on the environment. In the UK, there are strict rules to ensure that poisonous ammonia gas does not escape into the air or water.

### Exam tip

✓ In the exam, you may be asked to give reasons for Haber process conditions.

### Questions

1 Explain what is meant by yield.

2 Explain the purpose of using iron in the Haber process.

3 Explain why a pressure of 200 atm is chosen for the Haber process.

4 Explain why a temperature of 450°C is chosen for the Haber process.

5 Suggest why unreacted gases are recycled in the Haber process.

↓ E

↓ C

↓ A*

## Learning objectives

After studying this topic, you should be able to:

✔ explain equilibrium reactions

▲ Reaction between hydrogen chloride and ammonia to form ammonium chloride

**A** What is a closed system?

**B** Explain the meaning of the term dynamic equilibrium.

## Key words

**closed system**, **dynamic equilibrium**

## Reversible reactions

What does the picture on the left show?

Ammonia gas and hydrogen chloride gas from the bottle stoppers are reacting together. The product of the reaction is ammonium chloride. It forms as fumes of white solid.

ammonia + hydrogen chloride → ammonium chloride
$$NH_3(g) + HCl(g) \rightarrow NH_4Cl(s)$$

The reaction happens in reverse, too. If you heat ammonium chloride, it decomposes:

$$NH_4Cl(s) \rightarrow NH_3(g) + HCl(g)$$

Because the reaction can go in both directions, it is reversible, like the Haber process reaction.

## Dynamic equilibrium

Ammonia and hydrogen chloride can also react in a **closed system**. The picture shows a closed system – no materials can enter or leave the beaker in which the reaction takes place.

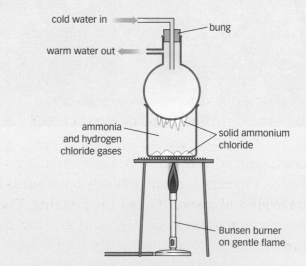

cold water in
bung
warm water out
ammonia and hydrogen chloride gases
solid ammonium chloride
Bunsen burner on gentle flame

▲ Hydrogen chloride and ammonia equilibrium apparatus

After a while, the system reaches **dynamic equilibrium**. The forward and backward reactions are both happening. The rate of reaction is the same in both directions. So the amount of each substance in the reaction mixture does not change.

$$NH_3(g) + HCl(g) \rightleftharpoons NH_4Cl(s)$$

## Another equilibrium reaction

If you heat calcium carbonate strongly in an open container, it decomposes. There are two products:
- carbon dioxide gas, which escapes
- calcium oxide, used for neutralising acid soils and making cement.

▲ Calcium carbonate decomposes on heating to make calcium oxide and carbon dioxide. Here, a man is heating limestone in an open kiln in Turkey to make calcium oxide.

Calcium carbonate also decomposes in a closed container. You need to pump the air out first, and then heat a few grams of the solid to about 800 °C. After a while, the amounts of substances in the container no longer change. Calcium carbonate is decomposing as fast as calcium oxide and carbon dioxide are reacting. The reactions occur at the same rate. Dynamic equilibrium has been reached.

$$\text{calcium carbonate} \rightleftharpoons \text{calcium oxide} + \text{carbon dioxide}$$
$$CaCO_3(s) \rightleftharpoons CaO(s) + CO_2(g)$$

### Did you know...?

Carbon dioxide in the atmosphere is in equilibrium with carbon dioxide dissolved in seawater.

$$\underset{\text{(atmosphere)}}{CO_2} \rightleftharpoons \underset{\substack{\text{(dissolved} \\ \text{in seawater)}}}{CO_2}$$

As more carbon dioxide enters the atmosphere from human activities, more carbon dioxide enters the oceans. The atmosphere–ocean system is not strictly a closed system.

### Exam tip

✔ Any reversible reaction that happens in a closed system will reach equilibrium.

### Questions

1 Write an equation to represent the dynamic equilibrium reaction of ammonia reacting with hydrogen chloride.

2 Explain why the amounts of substances in an equilibrium mixture do not change.

3 Suggest why companies that produce calcium oxide from limestone do the reaction in open kilns.

4 Describe how to establish a dynamic equilibrium between calcium carbonate, calcium oxide, and carbon dioxide.

↓
A*

## Learning objectives

After studying this topic, you should be able to:

- ✔ predict the effects of changing temperature on the amounts of substances in an equilibrium mixture.

**A** How does the chemist know when the reaction mixture has reached equilibrium?

**B** Predict what would happen if the chemist started by filling a glass tube with nitrogen dioxide, $NO_2$, instead of dinitrogen tetroxide, $N_2O_4$.

▲ The position of the equilibrium $N_2O_4(g) \rightleftharpoons 2NO_2(g)$ depends on the temperature

## Equilibrium

A chemist fills a glass tube with a colourless gas, dinitrogen tetroxide, $N_2O_4$. He seals the end of the tube, and waits. Gradually, a brown colour appears in the tube. The colour gets darker and darker, until it no longer changes. What's going on?

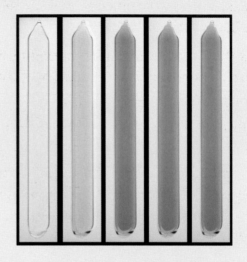

In the tube, dinitrogen tetroxide decomposes to form a brown gas, nitrogen dioxide, $NO_2$:

$$N_2O_4(g) \rightarrow 2NO_2(g)$$

As nitrogen dioxide forms, the reverse reaction starts to happen:

$$2NO_2(g) \rightarrow N_2O_4(g)$$

The brown colour gets darker until both reactions are happening at the same rate. The system is now in dynamic equilibrium.

$$N_2O_4(g) \rightleftharpoons 2NO_2(g)$$

## Changing temperature

The amounts of substances in an equilibrium mixture depend on the conditions of the reaction. The picture shows what happens if you heat and cool the $N_2O_4(g) \rightleftharpoons 2NO_2(g)$ equilibrium mixture:

- The tube on the left has been warmed. There is more brown $NO_2$ in the equilibrium mixture. Warming shifts the equilibrium $N_2O_4(g) \rightleftharpoons 2NO_2(g)$ to the right.
- The tube on the right has been cooled. There is more colourless $N_2O_4$ in the equilibrium mixture. Cooling shifts the equilibrium to the left.

## Explaining equilibrium shifts

Energy changes explain the different amounts of substances in the equilibrium mixture at different temperatures. For the equilibrium reaction below, the forward reaction is endothermic.

$$N_2O_4(g) \rightleftharpoons 2NO_2(g) \qquad \Delta H = +58 \text{ kJ/mol}$$

Increasing the temperature shifts the equilibrium to the right, to absorb the extra heat. The yield of $NO_2$ from the endothermic reaction increases.

Decreasing the temperature shifts the equilibrium to the left, to give out more heat. The yield of $N_2O_4$ from the exothermic reaction increases.

For the Haber process reaction, the forward reaction is exothermic.

$$N_2(g) + 3H_2(g) \rightleftharpoons 2NH_3(g) \qquad \Delta H = -92 \text{ kJ/mol}$$

At low temperatures, the position of the equilibrium shifts to the right. So the yield of ammonia is greater at low temperatures.

## Questions

1  Describe what happens to the yield of an exothermic equilibrium reaction as the temperature increases.

2  Describe what happens to the yield of an endothermic equilibrium reaction as the temperature decreases.

3  Use ideas about exothermic and endothermic reactions to explain why the yield of ammonia is greater at low temperatures.

↓
A*

4  The equation below represents an important step in the production of sulfuric acid:

$$2SO_2(g) + O_2(g) \rightleftharpoons 2SO_3(g) \qquad \Delta H = -197 \text{ kJ/mol}$$

Which would increase the yield of sulfur trioxide, $SO_3$ – increasing the temperature, or decreasing the temperature? Explain your answer.

## Vital acid

Do you enjoy relaxing in a bubbly bath, eating delicious food, or travelling by car? One important chemical helps make these activities possible – sulfuric acid. Every year, chemical plants all over the world produce over 150 million tonnes of the acid.

▲ Car batteries use sulfuric acid

▲ Sulfuric acid helps make paints

▲ Sulfuric acid helps make detergents

▲ Sulfuric acid helps make phosphate fertilisers

## Important equilibrium

Making sulfuric acid involves several stages, including this equilibrium reaction:

sulfur dioxide + oxygen ⇋ sulfur trioxide

$$2SO_2(g) + O_2(g) \rightleftharpoons 2SO_3(g) \qquad \Delta H = -197 \, \text{kJ/mol}$$

The reaction is exothermic. So lowering the temperature increases the yield of sulfur trioxide, $SO_3$. The reaction is very slow at low temperatures, so a temperature of around 450 °C is chosen as a compromise.

A Explain why sulfuric acid is manufactured in huge quantities.

B Explain why a temperature of 450 °C is chosen for the equilibrium reaction shown on the right.

## Pressure matters, too

Increasing the pressure also increases the yield of sulfur trioxide. This is because there are fewer molecules shown on the right of the symbol equation than on the left:

$$2SO_2(g) \; + \; O_2(g) \; \rightleftharpoons \; 2SO_3(g)$$

Number of molecules on left $= (2 + 1) = 3$
Number of molecules on right $= 2$

For all equilibrium gas reactions, an increase in pressure favours the reaction that produces the fewest molecules, as shown by the symbol equation for the reaction.

The Haber process reaction equation shows a total of four molecules on the left and two on the right:

$$N_2(g) \; + \; 3H_2(g) \; \rightleftharpoons \; 2NH_3(g)$$

So an increase in pressure shifts the equilibrium to the right, and increases the yield of ammonia. This explains why a high pressure is chosen for the Haber process.

### Did you know...?

Sulfuric acid is produced in the upper atmosphere of Venus. Light energy from the Sun helps carbon dioxide, sulfur dioxide, and water vapour to react together to make the acid.

## Questions

1 Copy and complete: For gaseous reactions, an increase in pressure favours the reaction that produces the _____ molecules, as shown by the symbol equation for the reaction.

2 Predict the effect of decreasing the pressure on an equilibrium reaction.

3 Predict whether a high or low pressure would maximise the yield of nitrogen monoxide, NO, in the reaction below. Give a reason for your prediction.

$$4NH_3(g) + 5O_2(g) \rightleftharpoons 4NO(g) + 6H_2O(g)$$

4 Predict the effect of increasing pressure on the equilibrium reaction below. Give a reason for your prediction.

$$H_2(g) + I_2(g) \rightleftharpoons 2HI(g)$$

$\downarrow$
A*

### Exam tip  AQA

✔ In equilibrium gas reactions, an increase in pressure will favour the reaction that produces the least number of molecules, as shown by the symbol equation for the reaction.

## Key words

ethanol, alcohol, organic compound, functional group, homologous series, propanol, methanol

## Depressant

Drugs alter normal bodily functions. Which recreational drug has the effects below?

*   slows reaction times
*   makes people forget things and feel confused
*   causes vomiting, unconsciousness, and even death.

The answer is **ethanol**, the **alcohol** in alcoholic drinks. In the UK, heavy drinking is blamed for up to 33000 deaths a year. A Swedish study found that up to 44% of deaths not caused by illness might be linked to ethanol, including those from falls, traffic accidents, suicide, and murder.

Drinking ethanol also has economic impacts. It makes huge profits for drinks companies. Its taxes bring income to the Government. But there are economic disadvantages, too. Ethanol costs the National Health Service an estimated £3 billion each year. The cost of policing alcohol-related crimes is also high.

> **A** List three social disadvantages of drinking alcohol.
>
> **B** List two economic advantages of alcohol, and two economic disadvantages.

## Inside alcohols

Ethanol is a compound of carbon, so it is an **organic compound**. Its molecules include the reactive –OH group. A reactive group of atoms in an organic molecule is called a **functional group**. Ethanol is not the only alcohol. It is a member of the **homologous series** of alcohols. The compounds of a homologous series have the same functional group, but a different number of carbon atoms. The table shows the first three members of the homologous series of alcohols.

| Name | Molecular formula | Structural formula |
|------|-------------------|--------------------|
| methanol | $CH_3OH$ | H—C—O—H (with H above and below C) |
| ethanol | $CH_3CH_2OH$ | H—C—C—O—H (with H atoms) |
| **propanol** | $CH_3CH_2CH_2OH$ | H—C—C—C—O—H (with H atoms) |

## Useful solvent

What's your favourite perfume or deodorant? Many perfumes and deodorants include ethanol as a solvent. Medicines, marker pens, and food flavourings may also include an ethanol solvent. **Methanol** makes a useful solvent, too.

Methanol and ethanol dissolve well in water. This is because the –OH group of their molecules is similar to water, $H_2O$, so alcohol and water molecules mix easily. Solutions of alcohols with water are neutral – their pH is 7.

◀ Deodorant often contains ethanol

### Questions

1 List three uses of ethanol.
2 What is an organic compound?
3 Draw the functional group in methanol, ethanol, and propanol.
4 Name the homologous series of which methanol is a member.
5 Draw up a table to summarise the economic and social advantages and disadvantages of the uses of alcohols.
6 Explain why ethanol dissolves well in water.

## Learning objectives

After studying this topic, you should be able to:

- ✔ describe some properties of alcohols
- ✔ describe and evaluate the use of alcohols as fuels

## Key words

**carbon neutral**

**A** Write a word equation for the combustion reaction of propanol.

**B** Write a balanced symbol equation for the combustion reaction of methanol.

## Alcohols burn

If you refuel your car in Brazil, you won't be adding just petrol to your tank. Brazilian car fuel is a mixture of petrol and ethanol, or even a mix of 95% ethanol with 5% water. In Brazil, ethanol is made by fermenting sugar cane.

Alcohols burn in air to make carbon dioxide and water. For example:

ethanol + oxygen → carbon dioxide + water

$CH_3CH_2OH(l)$ + $3O_2(g)$ → $2CO_2(g)$ + $3H_2O(g)$

▲ Drag racers often fuel their cars with methanol

## Pros and cons

There are advantages and disadvantages of using ethanol as a vehicle fuel instead of fossil fuels such as petrol and diesel.

| Advantages | Disadvantages |
|---|---|
| Made from a renewable resource, eg sugar cane or maize. | Crops from which ethanol is made are grown on land that could be used to grow food. |
| The plants from which the fuel is made remove carbon dioxide from the atmosphere are they grow. Some people say this means that ethanol fuel is **carbon neutral** – the plants remove the same amount of carbon dioxide from the atmosphere as burning the fuel later puts into the atmosphere. | Produces carbon dioxide, a greenhouse gas, as it burns. Carbon dioxide is also added to the atmosphere as a result of making fertilisers for the crops, and during the manufacture of ethanol from the crops. Some people say this means ethanol fuel is not carbon neutral. |

## Reaction with sodium

Alcohols react with sodium. The products are a salt and hydrogen. For example:

ethanol + sodium → sodium ethoxide + hydrogen

$$2CH_3CH_2OH(l) + 2Na(s) \rightarrow 2CH_3CH_2ONa(aq) + H_2(g)$$

H—C—C—O⁻Na⁺ structure:

$$H-\overset{\displaystyle H}{\underset{\displaystyle H}{C}}-\overset{\displaystyle H}{\underset{\displaystyle H}{C}}-O^-Na^+$$

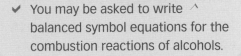
Water reacts with sodium more vigorously, but makes similar products:

water + sodium → sodium hydroxide + hydrogen

$$2H_2O(l) + 2Na(s) \rightarrow 2NaOH(aq) + H_2(g)$$

The reactions of sodium with ethanol and with water are similar because both water and ethanol contain an –OH group. During the reaction, O–H bonds break.

## Making vinegar

Francis leaves a bottle of wine open. A few days later, it tastes sour. The sour taste comes from ethanoic acid. Ethanoic acid forms when oxygen from the air oxidises some of the ethanol in the wine. Ethanoic acid is the main acid in vinegar.

You can also oxidise ethanol by
- the action of microbes
- reacting it with a chemical oxidising agent, such as acidified potassium dichromate(VI) solution.

▲ Orange acidified potassium dichromate(VI) solution oxidises ethanol to ethanoic acid. In the process, orange dichromate(VI) ions are reduced to green chromium ions, $Cr^{3+}$.

### Questions

1 Describe one advantage and one disadvantage of using ethanol as a fuel.

2 Name the products made when methanol burns.

3 Write a word equation for the reaction of propanol with sodium. The name of the salt formed is sodium propoxide.

4 List two ways by which ethanol can be oxidised to ethanoic acid.

5 Write a paragraph to evaluate the advantages and disadvantages of using ethanol made by the fermentation of sugar cane as a vehicle fuel.

↓ E

↓ C

↓ A*

## Learning objectives

After studying this topic, you should be able to:

✔ describe some properties and uses of carboxylic acids

▲ Dogs are better than humans at detecting the smells of carboxylic acids in sweat

## Sweat

Everyone has their own unique smell. Why? The answer is in your sweat. Everyone has a slightly different mix of compounds in their sweat. So everyone smells slightly different.

## Homologous series

The compounds that give sweat its smell are **carboxylic acids**. The carboxylic acids make up a homologous series. The series includes methanoic acid, ethanoic acid, and propanoic acid. All carboxylic acid molecules have the functional group –COOH, in which the atoms are arranged like this:

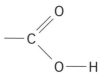

## Following a trend

As you go up any homologous series, there is a gradual change in some properties, such as boiling point. The boiling points and densities of some carboxylic acids are given in the table.

| Name of carboxylic acid | Molecular formula | Structural formula | Boiling point (°C) | Density (g/cm³) |
|---|---|---|---|---|
| methanoic acid | HCOOH | | 101 | 1.22 |
| ethanoic acid | $CH_3COOH$ | | 118 | 1.05 |
| propanoic acid | $CH_3CH_2COOH$ | | 141 | 0.99 |

> **A** Describe the trends in density and boiling point from methanoic acid to propanoic acid.
>
> **B** Draw the carboxylic acid functional group.

# Typical acids

Every carboxylic acid has the same functional group, so they all have similar chemical properties:

- They dissolve in water to make acidic solutions, with pH less than 7.
- They react with carbonates to produce carbon dioxide gas. For example:

ethanoic acid + calcium carbonate → calcium ethanoate + water + carbon dioxide

Calcium ethanoate is a salt.

# Using carboxylic acids

Carboxylic acids have many uses.

▲ Vinegar is a solution of ethanoic acid and other compounds in water

▲ Citric acid is added to many drinks to give them a sour taste

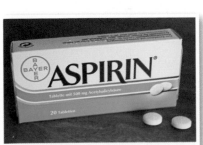

▲ Aspirin is an important carboxylic acid. It is used as a painkiller, and is also taken by people at risk of heart attack since it reduces blood clotting.

▲ Many fruits and vegetables contain ascorbic acid, vitamin C. Vitamin C is vital for health.

## Did you know...?

When an ant stings, it injects its victim with methanoic acid.

Butanoic acid gives rancid butter its foul smell, and vomit its sour taste.

## Exam tip

✔ You will not be expected to write balanced symbol equations for the reactions of carboxylic acids.

## Questions

1  Name three carboxylic acids.

2  Describe two uses of carboxylic acids.

3  Describe two properties of carboxylic acids that are typical of all acids.

4  Write a word equation for the reaction of propanoic acid with sodium carbonate. The salt formed is sodium propanoate.

5  Write a paragraph to describe the patterns in the properties of carboxylic acids.

## Learning objectives

After studying this topic, you should be able to:

✔ explain what makes carboxylic acids weak acids

## Key words

**ionise**, **strong acid**, **weak acid**

## Weak or strong?

It's the functional group that gives carboxylic acids their acid properties. When an ethanoic acid molecule dissolves in water, a hydrogen ion may leave its functional group. The acid molecule has split up to form ions. It has **ionised**.

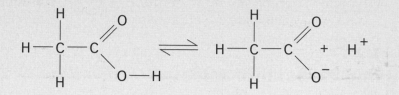

Hydrochloric acid is almost completely ionised in solution. Every hydrogen chloride molecule has split up to make hydrogen ions and chloride ions. The acid is fully ionised.

$$HCl(g) + (aq) \rightarrow H^+(aq) + Cl^-(aq)$$

Acids that are fully ionised in solution are **strong acids**. Hydrochloric acid, sulfuric acid, and nitric acid are all strong acids.

◀ Strong acids

> **A** Explain the meaning of the sentence: Hydrogen chloride is fully ionised in solution.

Carboxylic acids are different. Fewer than 1% of their molecules ionise when they dissolve in water. This means they are **weak acids**. The equilibrium for the solution of ethanoic acid lies towards the left:

$$CH_3COOH(l) + (aq) \rightleftharpoons CH_3COO^-(aq) + H^+(aq)$$

> **B** Explain what is meant by the term weak acid.

## Weak acids and pH

A solution is acidic if its pH is less than 7. The lower the pH, the more acidic the solution. pH is a measure of the concentration of hydrogen ions, $H^+$, in solution. The greater the concentration of hydrogen ions, the lower the pH.

Jack tests the pH of two acids of the same concentration. His results are in the table.

| Name and concentration of acid | pH |
|---|---|
| hydrochloric acid, 0.1 mol/dm³ | 1.0 |
| ethanoic acid, 0.1 mol/dm³ | 2.9 |

- The hydrochloric acid has a lower pH. It is more acidic. This is because all the hydrogen chloride molecules are ionised. The concentration of hydrogen ions is relatively high.
- The ethanoic acid has a higher pH. It is less acidic. This is because fewer than 1% of the ethanoic acid molecules are ionised. The concentration of hydrogen ions is relatively low.

▲ Universal indicator in a weak acid

## Questions

1. What is the difference between a strong acid and a weak acid?

2. Predict which will have the lower pH – a 1 mol/dm³ solution of ethanoic acid, or a 2 mol/dm³ solution of ethanoic acid.

3. Predict which will have the lower pH – a 1 mol/dm³ solution of propanoic acid or a 1 mol/dm³ solution of sulfuric acid.

4. Explain the difference between a weak acid and a dilute acid.

### Exam tip AQA

✔ There is a difference between a strong acid and a concentrated acid. Strong acids, such as hydrochloric acid, ionise completely in solution. A concentrated acid has a greater amount of acid dissolved in a certain volume of solution than a more dilute solution of the same acid.

## What gives an apple its smell?

Or a pear, or a strawberry, or an orange?

Complicated mixtures of chemicals give fruits their distinctive smells. But one group of compounds dominates fruit smells – the **esters**. Esters have the functional group –COO–, in which the atoms are arranged like this:

▲ The ester pentyl ethanoate contributes to the smell of pears

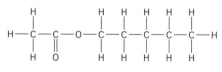

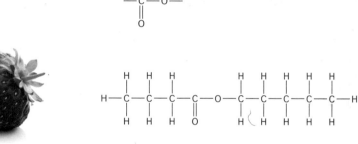

▲ The ester pentyl butanoate contributes to the smell of strawberries

### Did you know...?

Pear drop sweets have no pear in them – just the ester 3-methylbutyl ethanoate.

**A** Draw the functional group that appears in all esters.

**B** Give one property of esters.

## Ester properties

Most esters are liquids at room temperature. Liquid esters are **volatile** – they easily form vapours. And, of course, they all have distinctive smells.

## Using esters

The properties of esters determine their uses. Their smells and tastes make them useful for

- making sweet-smelling perfumes, shampoos, and shower gels
- flavouring foods, including sweets and chocolates.

The ester ethyl ethanoate is a useful solvent, since it is cheap, not poisonous, and has a pleasant smell. As a solvent, it is useful for cleaning circuit boards and removing caffeine from tea and coffee to make decaffeinated drinks.

# Making esters

Plants make a huge variety of natural esters. You can also make esters in the laboratory, by reacting a carboxylic acid with an alcohol. The diagrams show how to make ethyl ethanoate.

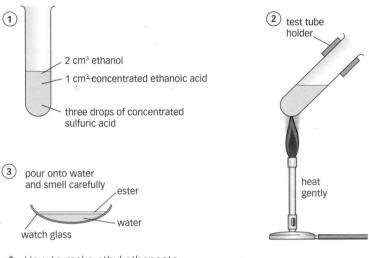

① 2 cm³ ethanol
1 cm³ concentrated ethanoic acid
three drops of concentrated sulfuric acid

② test tube holder
heat gently

③ pour onto water and smell carefully
ester
water
watch glass

▲ How to make ethyl ethanoate

The process of making an ester is called **esterification**. Concentrated sulfuric acid catalyses the reaction. The equation below summarises the reaction for making ethyl ethanoate from ethanol and ethanoic acid.

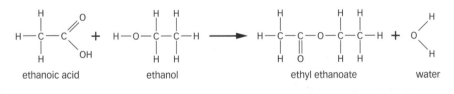

ethanoic acid     ethanol     ethyl ethanoate     water

## Key words

ester, volatile, esterification

## Questions

1 Explain the meaning of the term volatile.

2 What is a natural ester?

3 List three uses of esters, and explain how their properties make them suitable for these uses.

4 Name the reactants in the esterification reaction above.

5 Write a balanced symbol equation for the formation of ethyl ethanoate from ethanoic acid and ethanol. Use the equation that includes structural formulae to help you.

↓ E

↓ C

↓ A*

## Exam tip    AQA

✔ You could be given the formula of ethyl ethanoate and asked to name the ester. You also need to be able to recognise a compound as an ester from its name or structural formula. But you do not need to name – or draw structural formulae for – any ester other than ethyl ethanoate.

# Course catch-up

- Colours produced in flame tests are used to identify Group 1 ions. Precipitation reactions with sodium hydroxide identify other metal ions.

- Precipitation is used to identify Group 7 ions (halides) by reactions with silver nitrate, and to identify sulfate ions by reaction with barium chloride.

- Carbonate ions produce carbon dioxide when added to acids.

- In titrations, the volumes of acid and alkali reacting together are accurately measured. The results are used to calculate the concentration of solutions.

- Colour changes of indicators show the end point of titrations.

- Ammonia ($NH_3$) is manufactured when nitrogen and hydrogen are passed over an iron catalyst (Haber process). Nitrogen is obtained from air, and hydrogen from natural gas.

- The Haber process uses a temperature of 450°C and a pressure of 200 atmospheres. Conditions are chosen to make the yield as high as possible while keeping costs low. The Haber process is a reversible reaction.

- In closed systems, reversible reactions reach equilibrium when rate of forward reaction = rate of backward reaction.

- Increasing temperature makes the yield of an exothermic reaction lower.

- Increasing pressure makes the yield of a reaction higher if the number of gas molecules decreases.

- Alcohols (eg ethanol) are a homologous series of molecules containing the –OH functional group. They are used as fuels and solvents, and dissolve in water.

- Ethanol is oxidised to ethanoic acid. Ethanoic acid (found in vinegar) is a member of the carboxylic acid homologous series, containing the functional group –COOH.

- Carboxylic acids are weak acids and do not completely ionise to form H+ ions.

- Carboxylic acids have higher pH values than strong acids, and they react with carbonates to form carbon dioxide

- Carboxylic acids form esters when they react with alcohols in the presence of an acid catalyst.

- Esters (eg ethyl ethanoate) are pleasant-smelling compounds with the functional group –COO–. They are used in perfumes and flavourings.

titratio

acids react with alkalis

colour chang of indicator sh end point

volumes measured accurately

450°C, 200 atr iron catalyst

nitrogen from air and hydrogen from natural gas

conditio

MAKING AMMONIA

reversible reactions

but not completely ionised

rate of forward reaction = rate of backward reaction

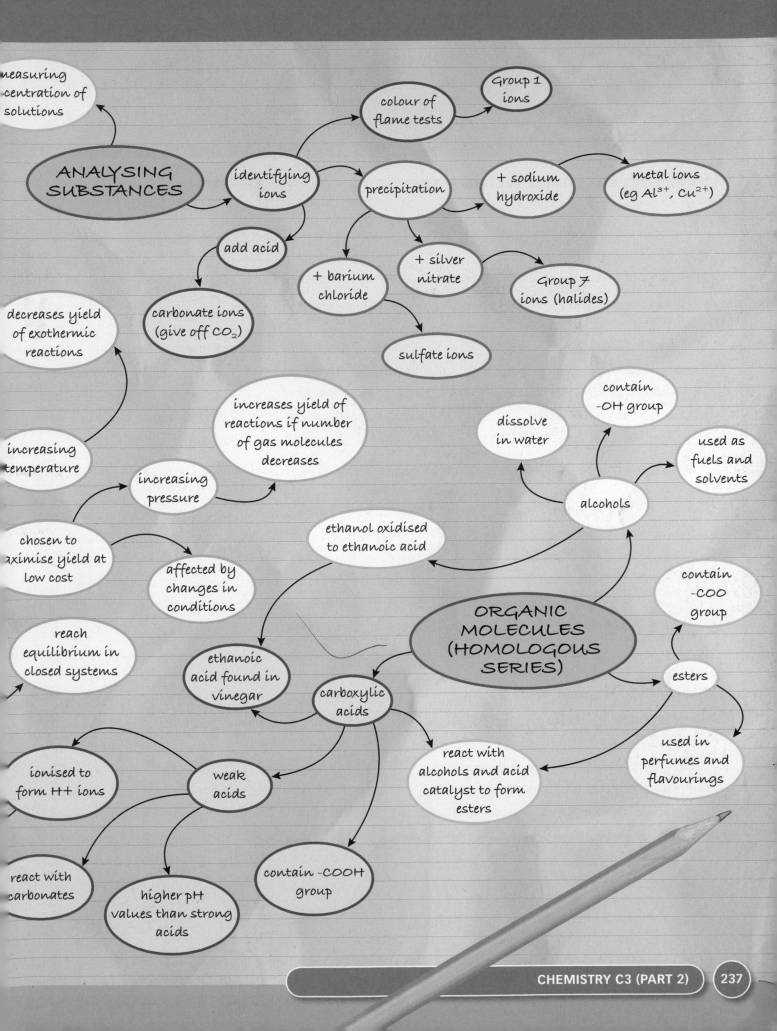

measuring
centration of
solutions

ANALYSING
SUBSTANCES

identifying
ions

colour of
flame tests

Group 1
ions

precipitation

+ sodium
hydroxide

metal ions
(eg $Al^{3+}$, $Cu^{2+}$)

add acid

+ barium
chloride

+ silver
nitrate

Group 7
ions (halides)

carbonate ions
(give off $CO_2$)

sulfate ions

decreases yield
of exothermic
reactions

increases yield of
reactions if number
of gas molecules
decreases

dissolve
in water

contain
-OH group

used as
fuels and
solvents

increasing
temperature

increasing
pressure

alcohols

ethanol oxidised
to ethanoic acid

chosen to
aximise yield at
low cost

affected by
changes in
conditions

ORGANIC
MOLECULES
(HOMOLOGOUS
SERIES)

contain
-COO
group

reach
equilibrium in
closed systems

ethanoic
acid found in
vinegar

carboxylic
acids

esters

react with
alcohols and acid
catalyst to form
esters

used in
perfumes and
flavourings

ionised to
form H+ ions

weak
acids

react with
carbonates

higher pH
values than strong
acids

contain -COOH
group

# Answering Extended Writing questions

**QUESTION**

Ammonia is manufactured using the Haber process. Describe how the Haber process is carried out and explain what conditions are used.

**The quality of written communication will be assessed in your answer to this question.**

**G–E**

You need nitrogen and iron to make ammonia. Getting a lot of ammonia is difficult because the reaction is revursable so special conditions are used.

Examiner: The answer mentions one raw material (nitrogen) but the second one is hydrogen (iron is a catalyst). The candidate has correctly explained the problem about the reversible reaction (note spelling). If the conditions had been listed, this might have been an answer in the D–C band.

**D–C**

Nitrogen and hydrogen from the air react to make ammonia using high temperature and an iron oxide catalyst. These help the reaction to be fast, the nitrogen and hydrogen are recycled.

Examiner: There is not much detail about the conditions – the candidate should give the actual temperature and mention that atmospheric pressure is used. The catalyst is actually iron and only nitrogen comes from the air. Spelling is fine, but a full stop should be used to split the last sentence into two.

**B–A\***

Nitrogen and hydrogen from are reacted together, but the reaction is in equilibrium. You would expect a high pressure and high temperature to make the yeald greater but that would be expensive so 450 °C and 200 atmospheres are used, also an iron catalyst. At the end the ammonia is cooled down which makes it a liquid.

Examiner: The candidate includes several correct facts about the conditions of the Haber process and goes some way to explaining why they are chosen. A full answer would also mention the need for increased rate. High temperature actually makes yield (note spelling) smaller, so 450 °C is chosen to give the optimum combination of rate and yield.

# Exam-style questions

**1**  Here are three organic molecules:

**A** ethanol

**B** ethyl ethanoate

**C** ethanoic acid

Complete the table with the correct letter, **A**, **B**, or **C**.

| | |
|---|---|
| An alcohol | |
| Found in vinegar | |
| Used as a fuel | |
| Formed when **A** and **C** react together | |
| Has a pH of less than 7 | |

**2** A chemical laboratory carries out tests on three substances.

| Test | D | E | F |
|---|---|---|---|
| Flame test | Lilac | Not done | Red |
| Reaction with sodium hydroxide | No reaction | Blue precipitate | White precipitate |
| Reaction with silver nitrate and nitric acid | Cream precipitate | White precipitate | Yellow precipitate |

Identify substances **D**, **E**, and **F**.

**3** William has a bottle of sodium hydroxide of unknown concentration. He takes 25 cm³ (0.025 dm³) of this solution and titrates it against hydrochloric acid with a concentration of 0.1 mol/dm³.

**a** 18.2 cm³ (0.0182 dm³) of hydrochloric acid was the average volume needed to neutralise the sodium hydroxide. The hydrochloric acid and sodium hydroxide react in a one-to-one ratio.

**i** Calculate the average amount in moles of hydrochloric acid used in the titration.

**ii** What amount in moles of sodium hydroxide reacts with this amount of hydrochloric acid?

**iii** Use your result from **ii** and other information from the question to calculate the concentration of the sodium hydroxide in mol/dm³.

**b** William thinks an unlabelled bottle of acid might be sulfuric acid, which contains sulfate ions. Describe how he could prove that it is sulfuric acid.

## Extended Writing

**4** Emily wants to do a titration to find out what volume of sulfuric acid is needed to neutralise a solution of sodium hydroxide. Write some instructions for a titration.

**5** Hydrochloric acid (HCl) is a strong acid and ethanoic acid ($CH_3COOH$) is a weak acid. Explain the meaning of these terms and describe how the properties of the two acids differ.

**6** The balanced equation for the Haber process (an exothermic reaction) is:

$$N_2 + 3H_2 \rightleftharpoons 2NH_3$$

A temperature of 450 °C and a pressure of 200 atm are used, along with an iron catalyst. Explain why.

A01  Recall the science

A02  Apply your knowledge

A03  Evaluate and analyse the evidence

# Glossary

**acid rain** Rain containing some gases from the air that make an acid solution. Acid rain dissolves some building materials such as limestone.

**activated carbon** Carbon that contains very many pores in its structure so has a very large surface area. This means that it can adsorb substances that would give water a bad taste or smell.

**activation energy** Minimum amount of energy that particles need in order to react when they collide.

**actual yield** Amount of product found by experiment.

**adsorption** Process in which molecules of a substance stick to the surface of a solid such as activated carbon.

**aggregate** Mixture of small stones that is mixed with cement and sand to make concrete.

**alcohol** Organic compound that contains the functional group –OH.

**alkali** Soluble base. Soluble hydroxides neutralise acids.

**alkali metal** An element in Group 1 of the periodic table (lithium, sodium, potassium, rubidium, caesium, francium).

**alkane** Hydrocarbon that contains no rings or double bonds in its molecule.

**alkene** Hydrocarbon that contains a double bond (two shared pairs of electrons) between two carbon atoms.

**alloy** Mixture of metals that has more useful properties than the metals on their own.

**alloy** Mixture of a metal with one or more other elements. The physical properties of an alloy are different from those of the elements in it.

**alternative fuel** Fuels other than fossil fuels, such as hydrogen or ethanol.

**ammonia** The simplest compound of hydrogen and nitrogen. The molecular formula is $NH_3$.

**anhydrous** Describes a salt that does not have its water of crystallisation.

**atmosphere** Mixture of gases surrounding the Earth.

**atom** The smallest particle of an element. Atoms have a nucleus, which contains protons and neutrons (except for hydrogen), and electrons, which are outside the nucleus.

**atomic number** The number of protons in the nucleus of an atom.

**balanced symbol equation** Equation that describes a chemical reaction and shows where all the atoms in the reactants appear in the products.

**base** Metal hydroxide or oxide. Bases neutralise acids.

**bauxite** The most common ore of aluminium, containing the mineral aluminium oxide.

**biofuel** Fuel obtained from a grown crop, such as ethanol from sugar cane.

**bioleaching** Process that uses bacteria to release compounds of metals from low-grade ores.

**blast furnace** The industrial structure in which iron is extracted from its ore.

**bond energy** The amount of energy required to break a bond in a molecule.

**breathable material** Material made up of several layers that will allow water vapour to pass through but not liquid water (rain).

**brine** Concentrated sodium chloride solution.

**bromine water** Solution of bromine in water. It is used as a test for unsaturated molecules, which decolourise it.

**buckminsterfullerene** Form of carbon consisting of molecules containing 60 carbon atoms joined together to form a hollow sphere.

**burette** Piece of apparatus used to measure accurately the volume of liquid delivered into a container.

**carbon monoxide** Poisonous gas that contains one atom of carbon and one atom of oxygen in each molecule, CO.

**carbon neutral** Describes a fuel that absorbs as much carbon dioxide when it is made as it gives out when it burns.

**carboxylic acid** Organic compound that contains the functional group –COOH.

**cast iron** Iron obtained by mixing melted iron from a blast furnace with scrap steel.

**catalyse** To speed up a reaction using a catalyst.

**catalyst** Substance that speeds up a reaction without being used up in the reaction.

**cement** Building material made by heating a mixture of powdered limestone and clay and then adding some calcium sulfate.

**chemical reaction** Process in which substances react to form different substances.

**chromatogram** The record obtained in chromatography.

**climate change** Change in long-term weather patterns.

**closed system** System in which no material can enter or leave.

**collide** Particles collide when they come into contact as one or both of them is moving.

**combustion** Process of burning fossil fuels, which releases carbon dioxide into the atmosphere and provides energy.

**complete combustion** The burning of a hydrocarbon to produce carbon dioxide and water only.

**compound** Substance made up of two or more different elements, chemically combined.

**concentrate** Ore that has been treated after mining to increase the percentage of metal in it.

**concentration** Mass of solute in a given volume of solution, usually measured in $g/cm^3$ or $mol/cm^3$.

**concrete** Building material made by mixing cement, sand, and aggregate with water.

**condense** Process of changing a gas into a liquid by lowering the temperature.

**continental drift** See *crustal movement*.

**core** Innermost layer of the earth, which is rich in nickel and iron.

**corrosion** The reaction of a metal with chemicals, especially water and oxygen, in the air. Corrosion weakens structures made from the metal.

**covalent bond** Shared pair of electrons that holds two atoms together.

**cracking** Process of breaking large molecules into smaller ones by heating or by using a catalyst.

**crude oil** Liquid mixture of hydrocarbons, formed by the decay of the remains of creatures that lived millions of years ago.

**crust** Outer, solid layer of the Earth.

**crustal movement** Movement of the tectonic plates caused by the slow flowing of the mantle.

**cryolite** Substance in which aluminium oxide dissolves in an aluminium electrolysis cell, so that the process can take place at a lower temperature than would be the case if pure aluminium oxide were used.

**crystallisation** Process by which solid crystals are obtained from solution.

**deforestation** Process of cutting down large numbers of trees to clear the ground for building or for growing crops.

**delocalised** Describes electrons that are free to move throughout a structure.

**dental polymer** Polymer that is used to fill holes in teeth as an alternative to fillings containing mercury.

**desalination** Process of removing dissolved salts from seawater, making the water fit to drink.

**diamond** Form of the element carbon in which each carbon atom is joined to four other carbon atoms by strong covalent bonds in a giant covalent structure.

**displacement reaction** Reaction in which a less reactive element is displaced from its compounds by a more reactive element.

**displayed formula** Description of a covalently bonded compound or element that uses symbols for atoms. It also shows the covalent bonds between the atoms.

**distillation** Process of obtaining a product from a mixture by boiling the mixture so that the product evaporates and can be collected.

**dot and cross diagram** Diagram that shows how the electrons are arranged in a molecule or in ions.

**double covalent bond** Strong bond between two atoms that consists of two shared pairs of electrons.

**dynamic equilibrium** The state of a reversible reaction when the forward and backward reactions are taking place at the same rate.

**earthquake** Shock caused by the sudden slipping of one tectonic plate past another.

**economic impact** The effect that the presence or absence of a material or activity has on the financial affairs of a community.

**electrode** Conductor through which an electric current enters or leaves a melted or dissolved ionic compound in electrolysis.

**electrolysis** Process by which melted or dissolved ionic compounds are broken down by passing an electric current through them.

**electron** Sub-atomic particle found outside the nucleus of an atom. It has a charge of –1 and a very small mass compared to protons and neutrons.

**electroplating** Covering an object with a thin coating of metal during an electrolysis reaction. The object to be plated is the negative electrode in an electrolysis cell.

**element** Substance that contains only atoms with the same atomic number.

**emulsifier** Material that enables a stable mixture to be formed between two liquids that would normally separate, such as oil and water.

**emulsion** Mixture containing small drops of one liquid spread throughout a second liquid.

**end point** The point in a titration when the chemical being added has exactly reacted with the chemical in the flask or beaker.

**endothermic** Describes a reaction in which heat is absorbed from the surroundings.

**endothermic reaction** Reaction in which energy is transferred from the surroundings to a reacting mixture.

**energy level** One of the shells that can be occupied by electrons in atoms.

**energy level diagram** Diagram showing the relative amounts of energy stored in the reactants and products of a reaction.

**E-number** Number used to indicate that a food additive has been approved by the European Union.

**environmental impact** The effect that the presence of a material or activity has on the local ecosystem.

**ester** Compound formed by the reaction of an alcohol with an acid. Their molecules contain the functional group –COO–.

**esterification** Reversible reaction between an acid and an alcohol to form an ester.

**ethanoic acid** Carboxylic acid with the molecular formula $CH_3COOH$.

**ethanol** Alcohol whose molecular formula is $CH_3CH_2OH$.

**evaporate** Process by which molecules leave a liquid and form a gas.

**exothermic** Describes a reaction in which heat is transferred to the surroundings.

**exothermic reaction** Reaction in which energy is transferred to the surroundings, in the form of heat, light, and sound, for example.

**fermentation** Process in which sugar is converted to ethanol by the action of enzymes in yeast.

**finite** Describes a resource, such as a fuel, that exists in a certain amount and that is not being replaced as it is used, so will eventually run out.

**flame test** Test that helps to identify metals in compounds by observing the colour a compound produces in the flame of a Bunsen burner.

**fluoridated water** Water that has had a fluorine compound (usually sodium fluoride) added to it. This helps to reduce tooth decay.

**focus** Point in the Earth's crust where tectonic plates slip and so cause an earthquake.

**formula** Description of a compound or an element that uses symbols for atoms. It shows how many atoms of each type are in the substance.

**fossil fuel** Fuel that was formed by the decay of the remains of creatures that lived millions of years ago. Coal, oil, and natural gas are all fossil fuels.

**fraction** Part of a mixture that can be separated from the rest by a physical process such as boiling.

**fractional distillation** Separation of a mixture into fractions that boil at different temperatures.

**fractioning column** Equipment used in fractional distillation to separate the fractions.

**fuel cell** Device that converts chemical energy directly to electrical energy without using combustion.

**fullerene** Form of the element carbon based on hexagonal rings of carbon atoms.

**functional group** Reactive group of atoms in an organic molecule that gives the molecule particular chemical properties.

**gas chromatography** Method of separating and identifying a mixture of substances, and comparing the amounts of the different substances in the mixture. Helium or nitrogen gas carries chemicals from a sample through a chromatography column.

**general formula** Formula that shows the relative number of atoms of the elements present in a molecule of any member of a homologous series such as the alkanes.

**giant covalent structure** Three-dimensional network of atoms that are joined together by covalent bonds in a regular pattern.

**giant ionic lattice** Regular, three-dimensional pattern of ions held together by strong electrostatic forces.

**giant metallic structure** Regular, three-dimensional pattern of positively charged metal ions held together by a sea of delocalised electrons.

**global dimming** The effect caused by particulates in the atmosphere reflecting sunlight back into space so that less light reaches the Earth's surface.

**global warming** Effect caused by greenhouse gases building up in the atmosphere and causing a rise in the global temperature.

**glucose** Sugar with the formula $C_6H_{12}O_6$.

**graphite** Form of the element carbon in which each carbon atom is joined to three other carbon atoms by strong covalent bonds to form a layer of carbon atoms. There are many of these layers in a piece of graphite. The layers are held together by delocalised electrons that move between the layers.

**greenhouse gas** Gas in the atmosphere that reduces the amount of heat lost to space by the Earth.

**group** Column in the periodic table that contains elements with similar properties.

**Haber process** Industrial process by which ammonia is made from hydrogen and nitrogen gases.

**half equation** Equation that describes the loss or gain of electrons by a reactant atom or ion, but is not balanced overall.

**halide ion** Negative ion formed from a halogen atom by the loss of one electron ($F^-$, $Cl^-$, $Br^-$, $I^-$, $At^-$).

**halogen** An element in Group 7 of the periodic table (fluorine, chlorine, bromine, iodine, astatine).

**hard water** Water that does not easily form a lather with soap.

**HDPE** Form of poly(ethene) (high density poly(ethene)) whose molecules do not contain side branches so that they can pack closely together. HDPE is rigid and not transparent.

**heat exchanger** Device for transferring energy from a hot output stream to a cold input stream so that the energy is not wasted.

**high carbon steel** Steel that has a relatively high carbon content (between 0.6 and 1.0%).

**high density poly(ethene)** Type of poly(ethene) that is rigid and not transparent. The polymer molecules can pack closely together because they do not contain side branches.

**homologous series** Series of organic compounds that contain different numbers of carbon atoms but the same functional group. They have a common general formula.

**hydrated** Describes a salt that contains water of crystallisation in its structure, such as $CuSO_4.5H_2O$.

**hydration** The addition of water to a substance in a chemical reaction to make a new compound.

**hydrocarbon** Compound whose molecules contain atoms of hydrogen and carbon only.

**hydrogel** Polymer that is able to absorb hundreds of times its own weight of water.

**hydrogen ion** Hydrogen atom that has lost its electron, $H^+$.

**hydrogenation** Process of combining an unsaturated molecule with hydrogen to remove the double bonds.

**hydrophilic** Describes the part of a molecule that interacts well with water.

**hydrophobic** Describes the part of a molecule that does not interact well with water but does interact well with oil.

**hydroxide ion** Ion with the formula $OH^-$. It makes solutions alkaline.

**incomplete combustion** The burning of a hydrocarbon to produce substances that could be oxidised further, such as carbon (soot) and carbon monoxide.

**intermolecular forces** Relatively weak forces between a molecule and its neighbours.

**internal combustion engine** Engine in which the energy produced by burning fuel in a cylinder is used to drive a piston or turbine.

**ion** Atom or group of atoms that has gained or lost electrons, so is electrically charged.

**ion exchange column** Column containing an ion exchange resin. The resin holds sodium ions and as hard water passes over it the calcium or magnesium ions in the water are exchanged for sodium ions.

**ionic bond** Strong electrostatic force of attraction between oppositely charged ions. Ionic bonds act in all directions.

**ionic bonding** Strong electrostatic forces between oppositely charged ions. Ionic bonding acts in all directions.

**ionic compound** Compound made up of positively and negatively charged ions.

**ionic equation** Equation that summarises a reaction between ions by showing only the ions that take part in the reaction.

**ionise** Process in which a molecule splits into oppositely charged ions when it dissolves in water.

**isotope** Atoms of the same element that have different numbers of neutrons are called isotopes.

**joule** Unit of energy.

**kilojoule** An amount of energy equal to 1000 joules.

**landfill site** Large hole in the ground in which rubbish is dumped.

**law of octaves** Observation by the English chemist John Newlands (1837–1898) that when the elements are arranged in order of their atomic weight, each successive eighth element shows similar properties.

**LDPE** Form of poly(ethene) (low density poly(ethene)) whose molecules contain side branches so that they do not pack closely together. LDPE is flexible and transparent.

**lime cycle** The series of reactions in which calcium carbonate is converted to calcium oxide by heating, to calcium hydroxide by adding water, and back to calcium carbonate as the calcium hydroxide absorbs carbon dioxide from the air.

**limestone** Rock that consists mainly of calcium carbonate. Much limestone contains the shells of marine animals that lived in prehistoric times.

**limewater** Solution of calcium hydroxide in water.

**low carbon steel** Steel that contains very little carbon (less than 0.3%).

**low density poly(ethene)** Type of poly(ethene) that is flexible and transparent. The polymer molecules contain side branches that prevent them packing closely together.

**low-grade ore** Ore that contains only a small percentage of the desired metal.

**lubricant** Substance that helps moving parts slide over each other easily.

**macromolecule** Three-dimensional network of atoms that are joined together by covalent bonds in a regular pattern.

**mantle** Thick solid layer of the Earth between the core and the crust. Although it is solid it can flow very slowly.

**mass number** The total number of protons and neutrons in the nucleus of an atom.

**mass spectrometer** Instrument that identifies substances quickly and accurately, and that can give their relative molecular masses.

**maximum theoretical yield** Amount of product, calculated from the reaction equation, that would be obtained if none was lost when the reaction was carried out.

**metal** Element that conducts electricity and heat and can be drawn into wires and beaten into sheets. Metals form positive ions in ionic compounds with non-metals.

**metal halide** A compound made up of a metal and a halogen.

**metallic bonding** Forces of attraction between positively charged metal ions and delocalised electrons in a metal.

**methanoic acid** Carboxylic acid with the molecular formula HCOOH.

**methanol** Alcohol with the molecular formula $CH_3OH$.

**microstrainer** A rotating drum covered in a fine mesh, used to remove particles such as algae from water.

**Miller–Urey experiment** An experiment carried out in 1953 in which sparks were passed through a mixture of gases that was believed to be like the mixture that existed in the atmosphere in the early history of the Earth.

**mineral** Naturally occurring material that contains compounds of metals.

**mixture** Substance that contains more than one type of molecule not chemically joined together.

**mobile phase** In chromatography, the solvent that carries chemicals from a sample through a chromatography column or along a piece of chromatography paper.

**mole** The relative formula mass of a substance in grams.

**molecular formula** Description of a compound or an element that uses symbols for atoms. It shows the relative number of atoms of each type in the substance.

**molecular ion peak** Peak on a mass spectrograph whose mass represents the relative formula mass of the molecule being analysed.

**molecule** Particle made up of two or more atoms chemically bonded together.

**monomer** Material whose molecules are the individual structural units of a polymer.

**monounsaturated** Describes a substance whose molecules contain only one double carbon–carbon bond.

**mortar** Building material made by mixing cement and sand with water.

**nanoparticle** Tiny particle made up of only a few hundred atoms. It measures between 1 nanometre and 100 nanometres across.

**nanoscience** Study of nanoparticles.

**nanotube** Sheets of carbon atoms arranged in hexagons that are wrapped around each other to form a cylinder with a hollow core.

**natural catalyst** Catalyst such as an enzyme that occurs naturally and does not have to be made in a laboratory.

**neutralisation reaction** Reaction in which hydrogen ions react with hydroxide ions to produce water.

**neutron** Sub-atomic particle found in the nucleus of an atom. It has no charge and a relative mass of 1.

**noble gas** An element with eight electrons in the outer shell (highest occupied energy level) of its atoms. The noble gases make up Group 0 in the periodic table.

**non-biodegradable** Describes a material that is not decomposed by the action of bacteria.

**non-renewable** Describes a resource, such as a fuel, that was made a long time ago and is being used up faster than it is being made now.

**nucleus** The relatively heavy central part of an atom, made up of protons and neutrons.

**nutrient** Material that provides one or more chemicals needed by living things.

**open-cast mine** Mine in which the mineral is obtained by digging from the surface instead of drilling a shaft into the ground.

**ore** Mineral that contains so much of a metal that the metal can be extracted from it at a profit.

**organic compound** Compound of carbon.

**oxidation** Process in which oxygen is added to an element or compound, or an atom or ion loses electrons.

**oxide ion** An atom of oxygen that has gained two electrons to form an ion with two negative charges, $O^{2-}$.

**paper chromatography** Method of separating substances in a mixture by allowing a solution of the mixture to flow along a sheet of special paper. If the substances in the mixture travel at different speeds, they will be separated.

**partial combustion** See *incomplete combustion*.

**particulates** Small particles, mostly of carbon (soot), formed by the incomplete combustion of fossil fuels.

**percentage by mass** The percentage by mass of an element in a compound is equal to the number of grams of the element in 100 grams of the compound.

**percentage yield** Actual yield × 100 ÷ maximum theoretical yield.

**periodic table** Table in which the elements are arranged in rows (periods) and columns (groups) in order of their atomic number.

**permanent hard water** Water from which the hardness cannot be removed by boiling the water.

**pH scale** Scale that describes the concentration of hydrogen ions in a solution.

**phytomining** Process in which plants are used to absorb metals from low-grade ores to obtain a material that has a higher percentage of the metal than the original ore.

**pipette** Piece of apparatus designed to deliver accurately a particular volume of liquid.

**plant oils** Oils obtained from plants by crushing and pressing or by dissolving the oil in a solvent.

**plate tectonics** Theory that the Earth's crust is made up of about a dozen plates that move in response to the flow of the mantle below them.

**poly(ethene)** Polymer whose molecules are made up of large numbers of ethene molecules linked together by carbon–carbon bonds.

**poly(propene)** Polymer whose molecules are made up of large numbers of propene molecules linked together by carbon–carbon bonds.

**polymer** Material whose molecules are made up of many repeated units.

**polymerisation reaction** Reaction in which monomer molecules join together to make a polymer.

**polyunsaturated** Describes a substance whose molecules contain more than one double carbon–carbon bond.

**precipitate** Suspension of small solid particles, spread throughout a liquid or solution.

**precipitation reaction** Reaction in which a precipitate forms.

**pressing** Process of squeezing a substance (usually a plant material) to obtain a product such as an oil.

**primordial soup theory** Theory that life on Earth originated when molecules in the atmosphere reacted together in the presence of sunlight.

**product** Chemical that is produced during a chemical reaction.

**propanoic acid** Carboxylic acid with the molecular formula $CH_3CH_2COOH$.

**propanol** Alcohol with the molecular formula $CH_3CH_2CH_2OH$.

**proton** Sub-atomic particle found in the nucleus of an atom. It has a charge of +1 and a relative mass of 1.

**rate of reaction** Amount of product made ÷ time, or the amount of reactant used ÷ time.

**reactant** Chemical that is used up during a chemical reaction.

**reactivity series** List of metals in order of their reactivity. The most reactive metals are at the top of the series.

**reduction** Process in which oxygen is removed from a compound, or an atom or ion gains electrons.

**reinforced concrete** Cement that is made stronger by allowing it to set round steel bars or mesh.

**relative atomic mass** The relative atomic mass of an element compares the mass of atoms of the element with the mass of atoms of the $^{12}C$ isotope. It is an average of the values for the isotopes of the element, taking into account their relative amounts. Its symbol is $A_r$.

**relative formula mass** The relative formula mass of a substance is the mass of a formula unit of that substance compared to the mass of a $^{12}C$ carbon atom. It is worked out by adding together all the $A_r$ values for the atoms in the formula. Its symbol is $M_r$.

**reservoir** Store of material such as water or carbon.

**retention time** In chromatography, the time it takes for a chemical in a mixture to move through the stationary phase.

**reversible reaction** Reaction in which the products of a reaction can react to produce the original reactants.

**rough titration** Titration carried out to give an approximate value for the end point.

**salt** Compound that contains metal ions and that can be made from an acid.

**sand filter** A deep bed of sand that removes bacteria and fine particles of solid from water as the water passes through it.

**saturated** Describes a compound that contains no double bonds in its molecules.

**sedimentary rock** Rock such as limestone that was formed when the remains of sea creatures fell to the floor of the oceans and built up into a thick layer.

**shape memory alloy** Alloy that, when bent or twisted, keeps its new shape, but when heated will return to its original shape.

**shape memory polymer** Polymer whose molecules can return to their coiled state when they are heated after they have been melted, stretched, and cooled.

**smart material** Material that can change its properties according to its environment.

**smelting** Process of obtaining a metal from an ore by mixing it with other substances and heating it in a furnace.

**social impact** The effect that the presence of a material, such as a plastic, or an activity, such as quarrying, has on the lives of people who live in the community.

**soft water** Water that easily forms a lather with soap.

**soluble** Substance that dissolves in a given solvent.

**solute** Substance dissolved in a solvent to make a solution.

**specific heat capacity** The amount of energy required to raise the temperature of 1 g of a substance by 1 °C.

**stainless steel** Alloy of steel containing other elements, especially chromium. Stainless steel resists corrosion, so does not rust.

**state symbols** The symbols used to indicate the state of a substance in an equation (g, gas; l, liquid; s, solid; aq, dissolved in water).

**stationary phase** In chromatography, the medium through which the mobile phase passes.

**steam distillation** Process of passing steam through a liquid mixture to carry away a desired product, which is then condensed.

**steel** Alloy of iron made by removing carbon from cast iron and adding other metals, such as chromium.

**sterilise** Process of killing bacteria in water using the gases chlorine or ozone.

**strong acid** Describes an acid that has all of its molecules ionised when dissolved in water.

**sub-atomic particle** Describes particles from which atoms are made, including protons, neutrons, and electrons.

**successful collision** Collision between reactant particles that results in a reaction.

**sulfur dioxide** Compound that contains one sulfur atom and two oxygen atoms in each molecule, $SO_2$. It is usually present in acid rain.

**surface area** In a chemical reaction involving a solid, the area of a solid that is in contact with the other reactants.

**symbol** The letter or letters that chemists use to represent one atom of an element, eg the symbol for an atom of iron is Fe.

**tectonic plate** One of the large pieces of the Earth's crust that gradually moves across the surface of the mantle.

**temporary hard water** Water from which the hardness can be removed by boiling the water.

**thermal decomposition** The breaking down of a substance by the effect of heat.

**thermosetting polymer** Polymer that does not melt.

**thermosoftening polymer** Polymer that softens easily on heating and that can be moulded into new shapes.

**titration** Reaction carried out using a solution whose concentration is known to find out the concentration of a chemical in a second solution.

**transition element** Describes a metallic element in the central block of the periodic table. Transition elements are hard and have high melting points. They can form ions with different positive charges and are often good catalysts.

**transition metal** Metal such as titanium that is in the central block of the periodic table.

**unreactive** Describes an element or compound that does not readily take part in chemical reactions.

**unsaturated** Describes a substance whose molecules contain at least one double carbon–carbon bond.

**unsaturated hydrocarbon** Hydrocarbon whose molecules include carbon atoms that share more than one pair of electrons.

**viscous** Describes a liquid that is difficult to pour.

**volatile** Describes a liquid that evaporates quickly.

**volcanic activity** Eruption of volcanoes, releasing large amounts of gases into the atmosphere. Volcanic activity occurs at the boundaries where tectonic plates meet.

**volcano** Place on the Earth's surface where a weakness in the crust allows molten rock (magma) to escape to the surface.

**washing soda** The compound sodium carbonate, which removes hardness when it is dissolved in hard water.

**weak acid** Describes an acid that has only some of its molecules ionised when dissolved in water.

**yield** Amount of the required product made in a reaction.

# Index

# Reference material

## Periodic table

Times of discovery

| | |
|---|---|
| ☐ before 1800 | ☐ 1900–1949 |
| ☐ 1800–1849 | ☐ 1949–1999 |
| ☐ 1849–1899 | |

Key:

relative atomic mass
**atomic number**
name
atomic (proton) number

Example:
1.0
**H**
hydrogen
1

| Group | 1 | 2 | | | | | | | | | | | 3 | 4 | 5 | 6 | 7 | 8 |
|---|---|---|---|---|---|---|---|---|---|---|---|---|---|---|---|---|---|---|
| | | | | | | | | | | | | | | | | | | 4 **He** helium 2 |
| Period 2 | 7 **Li** lithium 3 | 9 **Be** beryllium 4 | | | | | | | | | | | 11 **B** boron 5 | 12 **C** carbon 6 | 14 **N** nitrogen 7 | 16 **O** oxygen 8 | 19 **F** fluorine 9 | 20 **Ne** neon 10 |
| 3 | 23 **Na** sodium 11 | 24 **Mg** magnesium 12 | | | | | | | | | | | 27 **Al** aluminium 13 | 28 **Si** silicon 14 | 31 **P** phosphorus 15 | 32 **S** sulfur 16 | 35.5 **Cl** chlorine 17 | 40 **Ar** argon 18 |
| 4 | 39 **K** potassium 19 | 40 **Ca** calcium 20 | 45 **Sc** scandium 21 | 48 **Ti** titanium 22 | 51 **V** vanadium 23 | 52 **Cr** chromium 24 | 55 **Mn** manganese 25 | 56 **Fe** iron 26 | 59 **Co** cobalt 27 | 59 **Ni** nickel 28 | 63.5 **Cu** copper 29 | 65 **Zn** zinc 30 | 70 **Ga** gallium 31 | 73 **Ge** germanium 32 | 75 **As** arsenic 33 | 79 **Se** selenium 34 | 80 **Br** bromine 35 | 84 **Kr** krypton 36 |
| 5 | 85.5 **Rb** rubidium 37 | 88 **Sr** strontium 38 | 89 **Y** yttrium 39 | 91 **Zr** zirconium 40 | 93 **Nb** niobium 41 | 96 **Mo** molybdenum 42 | (98) **Tc** technetium 43 | 101 **Ru** ruthenium 44 | 103 **Rh** rhodium 45 | 106 **Pd** palladium 46 | 108 **Ag** silver 47 | 112 **Cd** cadmium 48 | 115 **In** indium 49 | 119 **Sn** tin 50 | 122 **Sb** antimony 51 | 128 **Te** tellurium 52 | 127 **I** iodine 53 | 131 **Xe** xenon 54 |
| 6 | 133 **Cs** caesium 55 | 137 **Ba** barium 56 | 139 **La*** lanthanum 57 | 178.5 **Hf** hafnium 72 | 181 **Ta** tantalum 73 | 184 **W** tungsten 74 | 186 **Re** rhenium 75 | 190 **Os** osmium 76 | 192 **Ir** iridium 77 | 195 **Pt** platinum 78 | 197 **Au** gold 79 | 201 **Hg** mercury 80 | 204 **Tl** thallium 81 | 207 **Pb** lead 82 | 209 **Bi** bismuth 83 | 210 **Po** polonium 84 | (210) **At** astatine 85 | 222 **Rn** radon 86 |
| 7 | (223) **Fr** francium 87 | (226) **Ra** radium 88 | (227) **Ac**# actinium 89 | (261) **Rf** rutherfordium 104 | (262) **Db** dubnium 105 | (266) **Sg** seaborgium 106 | (264) **Bh** bohrium 107 | (277) **Hs** hassium 108 | (268) **Mt** meitnerium 109 | (271) **Ds** darmstadtium 110 | (272) **Rg** roentgenium 111 | | | | | | | |

Elements with atomic numbers 112–116 have been reported but not fully authenticated

*58–71 Lanthanides

| 140 **Ce** cerium 58 | 141 **Pr** praseodymium 59 | 144 **Nd** neodymium 60 | (145) **Pm** promethium 61 | 150 **Sm** samarium 62 | 152 **Eu** europium 63 | 157 **Gd** gadolinium 64 | 159 **Tb** terbium 65 | 163 **Dy** dysprosium 66 | 165 **Ho** holmium 67 | 167 **Er** erbium 68 | 169 **Tm** thulium 69 | 173 **Yb** ytterbium 70 | 175 **Lu** lutetium 71 |
|---|---|---|---|---|---|---|---|---|---|---|---|---|---|

#90–103 Actinides

| 232 **Th** thorium 90 | 231 **Pa** protactinium 91 | 238 **U** uranium 92 | 237 **Np** neptunium 93 | 239 **Pu** plutonium 94 | 243 **Am** americium 95 | 247 **Cm** curium 96 | 247 **Bk** berkelium 97 | 252 **Cf** californium 98 | (252) **Es** einsteinium 99 | (257) **Fm** fermium 100 | (258) **Md** mendelevium 101 | (259) **No** nobelium 102 | (260) **Lr** lawrencium 103 |
|---|---|---|---|---|---|---|---|---|---|---|---|---|---|

## Reactivity series of metals

| | |
|---|---|
| Potassium | most reactive ↑ |
| Sodium | |
| Calcium | |
| Magnesium | |
| Aluminium | |
| *Carbon* | |
| Zinc | |
| Iron | |
| Tin | |
| Lead | |
| *Hydrogen* | |
| Copper | |
| Silver | |
| Gold | |
| Platinum | ↓ least reactive |

(elements in italics, though non-metals, have been included for comparison)

## Formulae of some common ions

| Name | Formula | Name | Formula |
|---|---|---|---|
| Hydrogen | $H^+$ | Chloride | $Cl^-$ |
| Sodium | $Na^+$ | Bromide | $Br^-$ |
| Silver | $Ag^+$ | Fluoride | $F^-$ |
| Potassium | $K^+$ | Iodide | $I^-$ |
| Lithium | $Li^+$ | Hydroxide | $OH^-$ |
| Ammonium | $NH_4^+$ | Nitrate | $NO_3^-$ |
| Barium | $Ba^{2+}$ | Oxide | $O^{2-}$ |
| Calcium | $Ca^{2+}$ | Sulfide | $S^{2-}$ |
| Copper(II) | $Cu^{2+}$ | Sulfate | $SO_4^{2-}$ |
| Magnesium | $Mg^{2+}$ | Carbonate | $CO_3^{2-}$ |
| Zinc | $Zn^{2+}$ | | |
| Lead | $Pb^{2+}$ | | |
| Iron(II) | $Fe^{2+}$ | | |
| Iron(III) | $Fe^{3+}$ | | |
| Aluminium | $Al^{3+}$ | | |

# Acknowledgements

The publisher and authors would like to thank the following for their permission to reproduce photographs and other copyright material:

**p8T** 81a/Alamy; **p8B** Chris Pearsall/Alamy; **p13** Thomas Harris/Istockphoto; **p14** Andraž Cerar/Shutterstock; **p17** Charles D. Winters/Science Photo Library; **p18T** Hannamariah/Shutterstock; **p18ML** Charles D. Winters/Science Photo Library; **p18MR** Charles D. Winters/Science Photo Library; **p18B** Viktor1/Shutterstock; **p20** Charles D. Winters/Science Photo Library; **p22** Jim Edds/Science Photo Library; **p23** Mikeledray/Shutterstock; **p24T** 1000 Words/Shutterstock; **p24B** Adam Hart-Davis/Science Photo Library; **p25** Dmitri Melnik/Shutterstock; **p26** Andrew Lambert Photography/Science Photo Library; **p27** Andre Maslennikov/Photolibrary; **p28TL** Typhoonski/Dreamstime; **p28TM** Mark William Richardson/Shutterstock; **p28TR** Stephen Inglis/Shutterstock; **p28M** Rui Ferreira/Shutterstock; **p28B** Prism68/Shutterstock; **p30M** Populous; **p30R** Fraser Gray/Rex Features; **p30L** Jarmund/Vigsn's AS Architects MNAL/Nils Petter Dale/nispe@datho.no; **p32M** PHB.cz (Richard Semik)/Shutterstock; **p32R** DenisNata/Shutterstock; **p32L** Yuriy Egorov/Dreamstime; **p33** Dmitry Rukhlenko/Shutterstock; **p34** Mark Schwettmann/Shutterstock; **p35** University Libre de Bruxelles; **p36R** Shutterstock; **p36L** Scott Camazine/Science Photo Library; **p37T** Oranhall/Dreamstime; **p37B** Ken Lucas, Visuals Unlimited/Science Photo Library; **p38L** Paul Rapson/Science Photo Library; **p38MR** Brian A Jackson/Shutterstock; **p38ML** PeJo/Shutterstock; **p38R** Valeria73/Shutterstock; **p40** Robert Brook/Science Photo Library; **p41** Olga Utlyakova/Shutterstock; **p42T** Rockongkoy/Shutterstock; **p42M** Skyscan/Science Photo Library; **p42B** PeJo/Shutterstock; **p43** INSADCO Photography/Photolibrary; **p45T** Antonio S./Shutterstock; **p45B** Typhoonski/Dreamstime; **p46R** Stacey Newman/Istockphoto; **p46L** Chris Hellier/Science Photo Library; **p53** Science Source USGS/Science Photo Library; **p54L** Paul Rapson/Science Photo Library; **p54ML** Kirsty Wigglesworth/AP Photo; **p54MR** Bomshtein/Shutterstock; **p54R** Lisa M. Robinson/Getty Images; **p56BL** Vadim Ponomarenko/Shutterstock; **p56BR** Prill Mediendesign & Fotografie/Istockphoto; **p56T** Rex Features; **p58** Jozsef Szasz-Fabian/Shutterstock; **p59** AJ Photo/Science Photo Library; **p60** Jo unruh/Istockphoto; **p61** Anastasia Pelikh/Istockphoto; **p62T** Syagci/Istockphoto; **p62B** Picsfive/Shutterstock; **p63** Robert Brook/Science Photo Library; **p64T** Everynight Images/Alamy; **p64B** Jim Varney/Science Photo Library; **p65** The Travel Library/Rex Features; **p66R** Lars Christensen/Shutterstock; **p66TL** Martin Bond/Photolibrary; **p66BL** Viktor1/Shutterstock; **p68LT** szarzynski/Shutterstock; **p68M** Tobik/Shutterstock; **p68R** Inga Spence/Visuals Unlimited/Getty Images; **p68LB** Adam Woolfitt/Photolibrary; **p70TL** Suzannah Skelton/Istockphoto; **p70TR** Hannu Viitanen/Dreamstime; **p70B** Lia Minou/Istockphoto; **p71** Piotr Marcinski/Shutterstock; **p72T** Kameel4u/Dreamstime; **p72B** Peter zijlstra/Shutterstock; **p74** Nanka (Kucherenko Olena)/Shutterstock; **p76** Canadian Press/Rex Features; **p77** Peter Menzel/Science Photo Library; **p79BL** Pixel Wave/Shutterstock; **p79TL** Dorling Kindersley/Getty Images; **p79R** AJ Photo/Science Photo Library; **p81T** Paul Topp/Dreamstime; **p81B** Dirk Wiersma/Science Photo Library; **p82** Dhoxax/Shutterstock; **p83** William Ervin/Science Photo Library; **p89** Dr Peter Harris/Science Photo Library; **p90TL** Tim Scrivener/Rex Features; **p90TR** Dmitri Melnik/Shutterstock; **p90BL** Martyn F. Chillmaid/Science Photo Library; **p90BR** Georgette Douwma/Science Photo Library; **p92** Springfield Gallery/Fotolia; **p94** Shane White/Shutterstock; **p95** John Chumack/Science Photo Library; **p96T** Hulton Archive/Stringer/Getty images; **p96B** Charles D. Winters/Science Photo Library; **p98** Martin Lovatt/Istockphoto; **p99** Tom Watkins/Rex Features; **p100** Charles D. Winters/Science Photo Library; **p101** Eliza Snow/Istockphoto; **p102R** Eye of Science/Science Photo Library; **p102L** Oleg Mitiukhin/Istockphoto; **p104R** Radius Images/Photolibrary; **p104L** David Hoffman Photo Library/Alamy; **p105** Ryan Balderas/Istockphoto; **p106** Kevin Britland/Shutterstock; **p107** Us Air Force/Science Photo Library; **p108R** Fertnig/Istockphoto; **p108L** Brent Danley; **p109** Tony Mcconnell/Science Photo Library; **p110L** Tyler Olson/Shutterstock; **p110R** Steve Meddle/Rex Features; **p112** Hank Morgan/Science Photo Library; **p114** Philippe Plailly/Science Photo Library; **p116** Peter Menzel/Science Photo Library; **p118** Tracy Hebden/Alamy; **p119** Simon Fraser/Science Photo Library; **p120** Ivanru/Dreamstime; **p121** Martyn F. Chillmaid/Science Photo Library; **p122L** Martyn F. Chillmaid/Science Photo Library; **p122M** Charles D. Winters/Science Photo Library; **p122R** Charles D. Winters/Science Photo Library; **p123** Andrew Lambert Photography/Science Photo Library; **p129** Javier Trueba/MSF/Science Photo Library; **p130L** Daniel Gale/Shutterstock; **p130R** Photos.com; **p132R** Cordelia Molloy/Science Photo Library; **p132L** Martyn F. Chillmaid/Science Photo Library; **p135** Charles D. Winters/Science Photo Library; **p136R** Photos.com; **p136L** Martyn F. Chillmaid/Science Photo Library; **p138** Charles D. Winters/Science Photo Library; **p139** Charles D. Winters/Science Photo Library; **p140BL** Khanh Trang/Istockphoto; **p140BM** Jenny Swanson/Istockphoto; **p140BR** Peter Elvidge/Istockphoto; **p140T** Ruslan Gilmanshin/Istockphoto; **p141L** Nick Free/Istockphoto; **p141R** Michael Durham/Getty; **p142L** Andrew Lambert Photography/Science Photo Library; **p142TR** Cordelia Molloy/Science Photo Library; **p142BR** Charles D. Winters/Science Photo Library; **p143T** Carolyn A. Mckeone/Science Photo Library; **p143B** pablo del rio sotelo/Istockphoto; **p144R** Dr Jeremy Burgess/Science Photo Library; **p144L** Science Photo Library; **p146** Marcus Leith, London. Image courtesy of Corvi-Mora, London; **p148** Sidney Moulds/Science Photo Library; **p150** Lawrence Migdale/Science Photo Library; **p151** Charles D. Winters/Science Photo Library; **p152TL** Alex Segre/Rex Features; **p152B** David Partington/Istockphoto; **p152TR** Mint Photography/Alamy; **p154T** Solent News & Photo Agency/Rex Features; **p154B** JW.Alker/Photolibrary; **p156R** Adam Hart-Davis/Science Photo Library; **p156TL** Joe Gough/Istockphoto; **p156BL** Cristian Lucaci/Istockphoto; **p159** James Holmes, Hays Chemicals/Science Photo Library; **p165** Eye Of Science/Science Photo Library; **p166T** Duncan Walker/Istockphoto; **p166B** Science Photo Library; **p167** Ria Novosti/Science Photo Library; **p168** US Library Of Congress/Science Photo Library; **p169** Dave Lo/Mindspark; **p170** Zack Seckler/Getty Images News/Getty Images; **p171** Martyn F. Chillmaid/Science Photo Library; **p172L** Andrew Brookes, National Physical Laboratory/Science Photo Library; **p172R** Jerry Mason/Science Photo Library; **p173TL** Charles D. Winters/Science Photo Library; **p173TR** Charles D. Winters/Science Photo Library; **p173B** Philip Evans/Rex Features; **p174L** Rex Features; **p174M** Steve Meddle/Rex Features; **p174R** Emmeline Watkins/Science Photo Library; **p175TL** Matt Meadows, Peter Arnold Inc./Science Photo Library; **p175M** Matt Meadows, Peter Arnold Inc./Science Photo Library; **p175TR** Andrew Lambert Photography/Science Photo Library; **p175BR** Martyn F. Chillmaid/Science Photo Library; **p175BL** Andrew Lambert Photography/Science Photo Library; **p176** Andrew Lambert Photography/Science Photo Library; **p178L** DCA Productions/Taxi/Getty Images; **p178R** Alexey Samoylenko/Istockphoto; **p182** StockLite/Shutterstock; **p183TL** Environment Agency, Thames Region and South West Water; **p183TR** Rachel Dewis/Istockphoto; **p183BL** Nickos/Istockphoto; **p183BR** Mauro Fermariello/Science Photo Library; **p185R** Dennis Kunkel Microscopy, Inc./Visuals Unlimited/Corbis; **p185L** Michael Blann/Photodisc/Getty Images; **p186** Nikada/Istockphoto; **p187L** Jochen Tack/Photolibrary; **p187R** James Steidl/Istockphoto; **p188R** Floortje/Istockphoto; **p188L** Fuat Kose/Istockphoto; **p190** Byronsdad/Istockphoto; **p192** Jeannot Olivet/Istockphoto; **p195** Bruce Mackie/Science Photo Library; **p196L** David McNew/Getty Images News/Getty Images; **p196R** David McNew/Getty Images News/Getty Images; **p203** Du Cane Medical Imaging Ltd/Science Photo Library; **p204T** Kenneth Sponsler/Istockphoto; **p204B** Andrew Lambert Photography/Science Photo Library; **p205** Andrew Lambert Photography/Science Photo Library; **p206** Marcus jones/Istockphoto; **p207** Andrew Lambert Photography/Science Photo Library; **p208R** Tek Image/Science Photo Library; **p208L** Keith/Custom Medical Stock Photo/Science Photo Library; **p209R** Charles D. Winters/Science Photo Library; **p209L** Charles D. Winters/Science Photo Library; **p210** Richard McGowan/Istockphoto; **p212** AJ Photo/Science Photo Library; **p213** Steve Horrell/Science Photo Library; **p214** Martyn F. Chillmaid/Science Photo Library; **p216TR** Peter Hendrie/Photographer's Choice/Getty Images; **p216TL** Tomas Bercic/Istockphoto; **p216B** Brian McEntire/Istockphoto; **p218** Ozturk Kemal Kayikci/Istockphoto; **p219** Ken Lucas, Visuals Unlimited/Science Photo Library; **p220** Andrew Lambert Photography/Science Photo Library; **p221** Dirk Wiersma/Science Photo Library; **p222R** Charles D. Winters/Science Photo Library; **p222L** Charles D. Winters/Science Photo Library; **p224TL** Tjanze/Istockphoto; **p224BR** David R. Frazier Photolibrary, Inc./Science Photo Library; **p224BL** Pixtal Images/Photolibrary; **p224TR** photoL/Istockphoto; **p226L** Alain Juteau/Istockphoto; **p226R** Marshall Turner/Dreamstime; **p227** Rex Features; **p228** Robert Young/Istockphoto; **p229** Andrew Lambert Photography/Science Photo Library; **p230** Alistair Scott/Dreamstime; **p231TL** Rex Features; **p231TR** a4stockphotos/Fotolia; **p231BL** Clynt Garnham Medical/Alamy; **p231BR** Yasonya/Fotolia; **p232** Andrew Lambert Photography/Science Photo Library; **p233** Martyn F. Chillmaid/Science Photo Library; **p234TR** Aleksejs Pivnenko/Fotolia; **p234ML** Melisback/Fotolia; **p234BR** UniqueLight/Shutterstock

Cover image courtesy of GUSTO IMAGES/SCIENCE PHOTO LIBRARY.

Illustrations by Wearset Ltd, Peter Bull Studios and Peter Stayte.

Although we have made every effort to trace and contact all copyright holders before publication this has not been possible in all cases. If notified, the publisher will rectify any errors or omissions at the earliest opportunity.

# OXFORD
UNIVERSITY PRESS

Great Clarendon Street, Oxford OX2 6DP

Oxford University Press is a department of the University of Oxford.
It furthers the University's objective of excellence in research,
scholarship, and education by publishing worldwide in

Oxford   New York

Auckland   Cape Town   Dar es Salaam   Hong Kong   Karachi
Kuala Lumpur   Madrid   Melbourne   Mexico City   Nairobi
New Delhi   Shanghai   Taipei   Toronto

With offices in
Argentina   Austria   Brazil   Chile   Czech Republic   France   Greece
Guatemala   Hungary   Italy   Japan   Poland   Portugal   Singapore
South Korea   Switzerland   Thailand   Turkey   Ukraine   Vietnam

Oxford is a registered trade mark of Oxford University Press
in the UK and in certain other countries.

British Library Cataloguing in Publication Data

Data available

ISBN 978-0-19-913603-2

10 9 8 7 6 5 4 3 2

Printed in Great Britain by Bell and Bain, Glasgow

Paper used in the production of this book is a natural, recyclable product
made from wood grown in sustainable forests. The manufacturing process
conforms to the environmental regulations of the country of origin.